METAZOA

METAZOA

ANIMAL LIFE AND
THE BIRTH OF THE MIND

PETER GODFREY-SMITH

FARRAR, STRAUS AND GIROUX

NEW YORK

Farrar, Straus and Giroux
120 Broadway, New York 10271

Library of Congress Cataloging-in-Publication Data
Names: Godfrey-Smith, Peter, author.
Title: Metazoa : animal life and the birth of the mind / Peter Godfrey-Smith.
Description: First edition. I New York : Farrar, Straus and Giroux, 2020. I
    Includes bibliographical references and index.
Identifiers: LCCN 2020027260 I ISBN 9780374207946 (hardcover)
Subjects: LCSH: Metazoa.
Classification: LCC QL45.2 .G63 2020 I DDC 590—dc23
LC record available at https://lccn.loc.gov/2020027260

Designed by Gretchen Achilles

Our books may be purchased in bulk for promotional, educational,
or business use. Please contact your local bookseller or the Macmillan Corporate
and Premium Sales Department at 1-800-221-7945, extension 5442, or by e-mail
at MacmillanSpecialMarkets@macmillan.com.

www.fsgbooks.com
www.twitter.com/fsgbooks • www.facebook.com/fsgbooks

1   3   5   7   9   10   8   6   4   2

*Dedicated to all who lost their lives
in the Australian bushfires of 2019–20,
and to the people who fought the fires*

And let me in this place movingly admonish you, ye ship-owners of Nantucket! Beware of enlisting in your vigilant fisheries any lad with lean brow and hollow eye; given to unseasonable meditativeness. . . . "Why, thou monkey," said a harpooneer to one of these lads, "we've been cruising now hard upon three years, and thou hast not raised a whale yet. Whales are scarce as hen's teeth whenever thou art up here." Perhaps they were; or perhaps there might have been shoals of them in the far horizon; but lulled into such an opium-like listlessness of vacant, unconscious reverie is this absent-minded youth by the blending cadence of waves with thoughts, that at last he loses his identity; takes the mystic ocean at his feet for the visible image of that deep, blue, bottomless soul, pervading mankind and nature; and every strange, half-seen, gliding, beautiful thing that eludes him; every dimly-discovered, up-rising fin of some undiscernible form, seems to him the em-bodiment of those elusive thoughts that only people the soul by continually flitting through it. In this enchanted mood, thy spirit ebbs away to whence it came; becomes diffused through time and space; like Wickliff's sprinkled Pantheistic ashes, forming at last a part of every shore the round globe over.

—HERMAN MELVILLE, *Moby-Dick; or, The Whale*

CONTENTS

CONTENTS

METAZOA

1

PROTOZOA

Down the Steps

You walk ten steps down on a stairway shaped from breakwater rocks straight into the water, which is flat and still, right at the top of the tide. Sound recedes with gravity and light fades to soft green as you dip beneath the surface. All you can hear is your breathing.

Soon you are in a sponge garden, in a jumble of shapes and colors. Some of the sponges have the form of bulbs or fans, growing upward from the seafloor. Others spread sideways over whatever they find, in an irregular encompassing layer. Amid the sponges are what look like ferns and flowers, and also *ascidians* (with a silent "c"), pale pink spout-like structures with enamel patterns inside. The spouts resemble the downward-curved air funnels on the decks of ships, though these spouts face in every direction. They are covered by all manner of tangled life, often so encrusted that they appear to be part of the physical landscape in which things live rather than organisms in their own right.

But the ascidians make small shifts, as if asleep and half sensing you as you pass. Occasionally, and always startling me a little, an ascidian body half-collapses in place and visibly expels the water held inside the animal, as if with a shrug and sigh. The landscape comes to life and makes its own comment as you go by.

Among the ascidians are anemones and soft corals. Some corals take the form of a cluster of tiny hands. Each hand has the regularity of a flower, but a flower that grasps at the water around it. They clench and slowly open again.

You are swimming through something like a forest, surrounded by life. But in a forest, most of what you encounter is the product of a different evolutionary path: the plant path. In the sponge garden, most of what you see are animals. Most of those animals (all except the sponges themselves) have nervous systems, electrified threads that stretch through the body. These bodies shift and sneeze, reach and hesitate. Some react abruptly as you arrive. Serpulid worms look like tufts of orange feather fixed to the reef, but the feathers are lined with eyes, and they vanish if you come too close. One can imagine being in a green forest, and finding the trees sneezing and coughing, reaching out hands, glimpsing you with invisible eyes.

This slow swim out from shore is showing you remnants and relatives of early forms of animal action. You are not swimming into the past—the sponge, ascidian, and coral are all present-day animals, products of the same span of evolutionary time that produced humans. You are not among ancestors but far-removed cousins, distant living kin. The garden around you is made of the topmost branches of a single family tree.

Farther out and under a ledge is a tangle of feelers and claws: a banded shrimp. Its body, partly transparent, is just a few inches long, but antennae and other appendages extend its presence at least three times as far. This animal is the first I've mentioned that might see you as an object, rather than responding to washes of light and looming masses. Then a bit farther still, on top of the reef, an octopus is stretched out like a cat—a very camouflaged cat—with several arms extended and others curled. This animal watches you, too, more overtly than the shrimp, raising its head in attention as you pass.

Matter, Life, and Mind

Something was dredged from the depths of the North Atlantic by HMS *Cyclops* in 1857. The sample looked like seafloor mud. It was preserved in alcohol and sent to the biologist T. H. Huxley.*

The sample was sent to Huxley not because it seemed especially unusual, but because of an interest, both scientific and practical, in seafloors at the time. The practical interest stemmed from the project of laying deep-sea telegraph cables. The first cable to span and send a message across the Atlantic was completed in 1858, though it lasted only three weeks, when the insulation failed and the signal-carrying current leaked away into the sea.

Huxley looked at the mud, noted some single-celled organisms and puzzling round bodies, and stored the sample away for about ten years.

He returned to it then with a better microscope. This time he saw discs and spheres of unknown origin, and also a slime-like substance, a "transparent gelatinous matter," surrounding

---

* This book has a lot of notes collected at the end that give references to sources and follow paths some distance further. They are organized there with a page reference and the phrase in the main text that the path leads from.

them. Huxley suggested that he had found a new kind of organism, of an exceptionally simple form. His cautious interpretation was that the discs and spheres were hard parts produced by the jelly-like matter itself, which was alive. Huxley named the new organism after Ernst Haeckel, a German biologist, illustrator, and philosopher. The new form of life was to be called *Bathybius Haeckelii*.

Haeckel was delighted with both the discovery and christening. He had been arguing that something like this must exist. Haeckel, like Huxley, was entirely convinced by Darwin's theory of evolution, unveiled in *On the Origin of Species* in 1859. Huxley and Haeckel were the leading advocates of Darwinism in their respective countries, England and Germany. Both were also eager to press on to questions that Darwin had been reluctant, beyond a few brief passages, to speculate about: the origin of life and the beginning of the evolutionary process. Did life arise just once on Earth, or several times? Haeckel was convinced that the spontaneous generation of life from inanimate materials was possible, and might be going on continually. He embraced *Bathybius* as a fundamental form of life, one that might cover large tracts of the deep seafloor; he saw it as a bridge or link between the realm of life and the realm of dead, inorganic matter.

The traditional conception of how life is organized, a picture in place since the ancient Greeks, recognized just two kinds of living things: animals and plants. Everything alive had to fall on one side or the other. When the Swedish botanist Carl Linnaeus devised a new scheme of classification in the eighteenth century, he installed plant and animal kingdoms alongside a third, inanimate realm, the "kingdom of rocks," or *Lapides*. This three-way distinction is still seen in the familiar question, "animal, vegetable, or mineral?"

By the time of Linnaeus, microscopic organisms had been observed, perhaps first in the 1670s by the Dutch draper Antonie van Leeuwenhoek, who made the most powerful of the early microscopes. Linnaeus included a fair number of tiny, microscopically observed organisms in his classification of beings, putting them in the category of "worms." (He concluded the tenth edition of his *Systema Naturae*, the edition that began the classification of animals as well as plants, with a group he called *Monas*: "body a mere point.")

As biology progressed, puzzle cases began to appear, especially at the microscopic scale. The tendency was to try to put them with either plants (algae) or animals (protozoa), on one side of the boundary or the other. But it was often hard to tell where some new creature belonged, and natural to feel that the standard classification was under strain.

In 1860, the British naturalist John Hogg argued that the sensible thing to do was to cease the shoehorning and add a fourth kingdom for the small organisms, increasingly recognized as single-celled, that are neither plants nor animals. These he called *Protoctista*, and he placed them in a *Regnum Primigenum*, or "primeval kingdom," that accompanied animals, plants, and minerals. (Hogg's term, *Protoctista*, was later shortened by Haeckel to the more modern *Protista*.) As Hogg saw it, the boundaries between the different living realms were vague, but the boundary between the mineral kingdom and the living was sharp.

The wrangling of categories I've described has so far been concerned with life, not with the mind. But life and the mind have long seemed linked somehow, even if their perceived relationship has not been stable. In the framework of Aristotle, developed over two millennia earlier, *soul* unifies the living and the mental. Soul, for Aristotle, is a kind of inner form that directs bodily activities, and it exists in different levels or grades in different living things. Plants take in nutrients to keep themselves alive—that shows a kind of soul. Animals do this and can also sense their surroundings and respond—that is another kind of soul. Humans can reason, in addition to the other two capacities, and so have a third kind. For Aristotle, even inanimate objects that lack souls also often behave in accordance with purposes or goals, tending toward their natural place.

The overthrow of Aristotle's picture in the seventeenth century's "Scientific Revolution" included a redrawing of these relationships. This involved a hardened conception of the physical—the assertion of a mechanical, push-pull view of matter with little or no role for purpose—and a lifting or etherealization of the soul. The soul, integral to all living nature in Aristotle, became a more rarified, intellectual affair. Souls may also be saved by divine will, permitting a kind of eternal life.

For René Descartes, an especially influential figure in this period, there is a sharp divide between the physical and the mental, and we humans are a combination of both; we are physical *and* mental beings. We succeed in being both because the two realms make contact in a small organ in our brains. This is Descartes's "dualism." Other animals, for Descartes, lack souls and are purely mechanical—a dog is without feeling, no matter what is done to it. The souls that make humans special are no longer present, even in faint forms, in animals and plants.

In the nineteenth century, the time of Darwin, Haeckel, and Huxley, advances in biology and other sciences made dualism of Descartes's kind look less and less viable. Darwin's work suggested a picture in which the divide between humans and other animals is not so sharp. Different forms of life along with different mental powers might arise through gradual processes of evolution, especially by adaptation to circumstances and the branchings that originate species. This should suffice to explain both bodies and minds—if you can get things started.

That was a big *if.* Haeckel, Huxley, and others approached this part of the problem as follows. They thought there must be a *stuff*, present in living things, that enables both life and the beginnings of a mind. This stuff would be physical, not supernatural, but quite unlike ordinary matter. If we could isolate it, you could pick up a spoonful of it, and in your spoon it would still be the special stuff. They called it "protoplasm."

This might seem an odd approach, but it was motivated in part by close inspection of cells and simple organisms. When people looked inside cells, it seemed that not enough organization was present—not enough parts were different from other parts—for cells to do what they are evidently able to do. What they saw seemed to be just a substance, transparent and soft. The English physiologist William Benjamin Carpenter, writing in 1862, marveled at what single-celled organisms could achieve: the "vital operations" that one sees "carried on by an elaborate apparatus" in an animal are instead brought about by "a little particle of apparently homogeneous jelly." The particle of jelly is seen "laying hold of its food without members, swallowing it without a mouth, digesting it without a stomach," and "moving from place to place without muscles." This led Huxley, and others, to think that it could not be an intricate organization of ordinary matter that explains living activity, but a different ingredient, one that was inherently alive: "organization is the result of life, not life the result of organization."

Against that background, *Bathybius* seemed extraordinarily promising. It appeared to be a pure sample of the stuff of life, stuff that perhaps arises spontaneously all the time, forming an ever-renewing deep-sea organic carpet. Further samples were examined. *Bathybius* obtained from the Bay of Biscay was described as being capable of movement. Other biologists were not so sure about this alleged primordial life-form, however, and the growing mass of speculation around it. How was *Bathybius* staying alive down there? What might it eat?

Then came the *Challenger* expedition—a four-year project organized by the Royal Society of London in the 1870s that took samples from hundreds of deep-sea sites around the world. The aim was the first comprehensive inventory of life in the deepest waters. The chief scientist on the expedition, Charles Wyville Thomson, was willing to work on the *Bathybius* question although he was wary of it. No fresh samples were found by the *Challenger*, and two scientists aboard the ship began, amid some tinkering, to suspect that *Bathybius* was not alive and not even close to it. With a series of experiments, they showed that *Bathybius* appeared to be nothing more than the product of a chemical reaction between seawater and the alcohol used to preserve samples, including Huxley's old sample from the HMS *Cyclops*.

Bathybius was dead. Huxley acknowledged his error immediately. Haeckel, more committed to *Bathybius* as a missing link, hung on, unfortunately, for nearly ten more years. But the bridge had failed.

Afterward, some people still held out hope for a bridge of roughly the same kind—a special substance that would link life with matter. But in the years that followed, views of that kind subsided. They were replaced in a slow process of discovery, a process that eventually made living activity no longer mysterious. The resulting explanation of life proceeded in exactly the way that Huxley and Haeckel could not countenance: in terms of the hidden organization of ordinary matter.

That matter is not "ordinary" in every sense, as we will see, but it is ordinary in its basic composition. Living systems are made of the same chemical elements that make up the rest of the universe, running according to physical principles that extend also into the inanimate realm. We don't presently know how life originated, but its origin is no longer a mystery of a kind that might make us believe that some extra substance generates the living world.

This has been the triumph of a *materialist* view of life—a view that permits no supernatural intrusions. It was also the triumph of a view that sees the physical world itself as unified in its basic constituents. Living activity is not explained in terms of a mysterious ingredient, but in terms of intricate structure on a tiny scale. That scale is almost inconceivable. To pick just one example, ribosomes are important parts of cells—the stations where protein molecules are assembled—with a rather complex structure of their own. But over 100 million ribosomes could fit on the period printed at the end of this sentence.

Life, then, has fallen into place. In the case of the mind, much less is resolved.

The Gap

From the late nineteenth century onward, with Darwin's revolution gathering steam, it seemed hard to maintain a dualist view of the mind like Descartes's. Dualism makes some sense within an overall picture that locates humans as a unique and special part of nature, close to God in some way. Then all the rest, alive or dead, can be purely material, while we have an added ingredient. An evolutionary perspective on humanity, one that sees continuities between ourselves and other animals, makes dualism difficult, though not impossible, to maintain. This motivates the attempt to develop a materialist view of the mind, one that explains thought, experience, and feeling in terms of physical and chemical processes. The fact that life itself succumbed to a materialist treatment of this kind is encouraging, but it is not clear how much it really helps; it's not clear what relationship the success of materialism in biology has to the puzzles of the mind.

Looking again at the history, we can distinguish two alternative paths that continue through to the present. Aristotle, as we saw, recognized several different grades of soul, linking plants, animals, and ourselves. What we call "mind" is viewed as a natural extension, or version, of living activity. Aristotle's view was not an evolutionary one, but it is not too hard to recast such a picture in evolutionary terms. The evolution of complex life naturally gives rise to the mind, through the growth of purposeful action and sensitivity to the environment.

Descartes, in contrast, saw life as one thing and mind as entirely another. There is no reason, in this second view, to think that progress in understanding life will make much of a difference to problems about the mind.

Over the last century or so, most views in this area have been materialist, but in one respect they have moved close to Descartes. From the mid-twentieth century onward, theorists shifted away from seeing close connections between the nature of life and the mind. This was encouraged by the advance of computers. Computer technology, as it developed from the middle decades of the last century, promised a different bridge between the mental and the physical, a bridge made of logic rather than life. The new mechanization of reasoning and memory—computation—seemed a better way forward. As artificial intelligence (AI) systems developed, some of them started to seem a bit intelligent, but there was little reason to think of them as *alive*. Animal bodies, it seemed, did not matter very much—they came to appear entirely optional, in fact. Software was the heart of the matter. The brain runs a program, and that program might run on other machines (or things other than machines) as well.

These years also saw a sharpening of the problem of mental and physical. "The mind" as puzzle was replaced by a more specific conundrum. The new view holds that some of the mind can be fairly readily explained in materialist terms, while another aspect is more resistant. The resistant side is subjective experience, or consciousness. Consider memory, for example. We might find that various kinds of animals have memory; they create traces of the past in their brains, and use those traces later, when working out what to do. It is not too hard to imagine how brains might achieve this. Much of that problem is unsolved, but it certainly looks soluble; we should be able to work out how this side of memory works. But in humans, at least, some kinds of memory also *feel* like something. As Thomas Nagel put it in 1974, there is *something it's like*—something it *feels* like—to have a mind. There is something it feels like to remember a good experience, or a bad one. The "information-processing" side of memory, the ability to store and retrieve useful information, might either be accompanied by this additional feature or not. The hard part of the mind-body problem is explaining that last side of our mental lives, explaining in biological, physical, or computer-based terms how felt experience can exist in the world.

This problem is often still approached through a range of classic options. The main divide has materialist (or "physical-ist") views on one side, and dualism on the other. More radical possibilities are also entertained. *Panpsychism* holds that all matter, including the matter in objects like tables, has a mental aspect to it. This is not the idea that the entire universe is made of experience—that is *idealism*. Instead, a panpsychist accepts the physical layout of the world as it appears, but adds that the material that makes up that world always has a side to it that is faintly mind-like. This mind-like side of matter gives rise to experience and consciousness, once some of that matter is organized into brains. Despite its apparent extravagance, panpsychism has serious defenders. Thomas Nagel, who I mentioned above, argues that panpsychism should be kept on the table as an option, because every view has significant problems and panpsychism's problems are no worse than others'. Ernst Haeckel, in the post-bathybius years, was also attracted to panpsychism. Huxley was attracted to another unorthodox view. He suspected that conscious experience might be an effect of material processes, but never a cause of them. This is an unusual kind of dualism, and it also has defenders today.

Something that is vivid in the wild sweep of these alternative views of the universe, and visible also in more mundane discussions, is a huge diversity in ideas about where minds are to be found. For some, mind is everywhere, or nearly everywhere. For others, it is confined to humans and perhaps a few animals similar to us. One person will look at a paramecium, a single-celled organism, swimming vigorously through a film of water and say: What is going on in that creature is enough for it to have feelings. The paramecium is responsive, and has goals. On a tiny scale, it has experience. Another person will not merely dismiss the paramecium, but will look at a complex animal, like a fish, and say: There is probably no feeling there at all. The fish has a lot of reflexes and instincts, and some fairly complicated brain activity, but all of this activity is going on "in the dark." If this second person is wrong, why are they wrong? If panpsychism is also wrong and there is no hint of feeling in a grain of sand, why is *that* wrong? *Might* things be that way? There often seems to be a kind of arbitrariness in the situation. People can say whatever they like. If I were to guess where most people stand at the moment, when asked which living things around them have experiences, I would conjecture that a common answer is "yes" for mammals and birds, "perhaps" for fish and reptiles, and "no" for everything else. But if someone insists on pushing further out (to ants, plants, and paramecia) or pulling further in (mammals only), the discussion rapidly gets a bit untethered. How could we possibly work out who is right?

This sense of arbitrariness is related to something the philosopher Joseph Levine has called "the explanatory gap." Even if we come to be pretty sure that the mind must have a purely physical basis, with nothing added, we would also want to know why *this* physical setup gives rise to *this* kind of experience, rather than something else. Why does it feel like *this* to have a brain of the particular kind you have, going through the processes it is going through right now? Even if the difficulties faced by other views convince us that materialism has to be true, it's hard to see *how* it's true, how things could be this way.

That is the cluster of problems I want to address in this book. The aim is not to answer Levine's questions about particular experiences—which activities of the brain are involved in seeing color or feeling pain. That is a task for neuroscience. The aim, instead, is to make sense of why it feels like something to be a material being of the kind that we are. That "we" is intended to be rather broad; my main target is not the intricacies of human consciousness, but experience in general, something that might extend to many other animals. I want to address these questions about experience in a way that reduces the sense of arbitrariness I described above—the feeling that you could say yes to bacteria, no to birds, whatever strikes you.

The approach I take to the mind-body problem is biological, and one that fits into a materialist picture of the world. The word "materialism" to many suggests a hard-headed, tough-minded view: the world is smaller than you thought, less special or less sacred, just atoms bumping into each other. Atoms bumping into each other are indeed quite important, but I do not want to get the story moving with a mood of toughness and restriction in the air. The "physical" or "material" world is more than a world of thudding collisions or dry structure. It is a world of energy and fields and hidden influences. We should be ready for ongoing surprises about what it contains.

The approach taken in this book is a biological materialism, but in many ways the heart of my outlook is a broader position, sometimes called *monism*. Monism is a commitment to an underlying unity in nature, a unity at the most basic levels. Materialism is one kind of monism, as it is committed to the idea that mental phenomena, including subjective experience, are manifestations of more basic activities described in biology, chemistry, and physics. Idealism, the idea that everything is mental, is another kind of monism—it is a different assertion of unity. (An idealist must explain how what seem to be physical objects and goings-on are really manifestations of mind or spirit.) Yet another way of being a monist is to think that both what we call the "physical" and what we call the "mental" are manifestations of something else that is basic; this view is called *neutral* monism. Rather than explaining the mental in physical terms or explaining the physical in mental terms, we explain both the physical and mental in terms of something else. That "something else" tends to remain rather mysterious. If I was not a materialist I'd be a neutral monist, but that is an outside possibility for me. The way I will proceed is by starting with life—understood in a materialist way—and trying to show how the evolutionary development of living systems can give rise to minds. I want to close—partially, at least—the explanatory gap between mental and physical.

Before we proceed, however, let's take a closer look at the mental side of the puzzle, and the words we use to describe it. The side of the mind that Nagel tried to point to by saying "there is something it's like . . ." is now often called *consciousness*. (Nagel himself calls it that.) You are conscious, in this sense, if there is something it feels like to be you. But the term "consciousness" is often misleading here, as it tends to suggest something quite sophisticated. That phrase "something it's like . . ." is supposed to include the presence of feelings of any kind. There is something it feels like to be you—or a fish, or a moth—if the vaguest, dimmest washes of sensation are part of your life. The fact that the word "consciousness" suggests more than this tends to cause trouble.

For example, neuroscientists often say that consciousness depends on the cerebral cortex, the folded part at the top of our brains, something found only in mammals and some other vertebrates. Here is a quote from the physician and essayist Oliver Sacks, talking about a patient who had, as a result of a brain infection, lost all ability to hold new events in memory. Sacks asked: "What is the relationship of action patterns and procedural memories, which are associated with relatively primitive portions of the nervous system, to consciousness and sensibility, which depend on the cerebral cortex?" Sacks is asking a question here, but also stating an assumption: that consciousness and sensibility depend on the cerebral cortex. Does Sacks mean that if someone or something lacks a cerebral cortex, they will lack consciousness in its here-I-am richness, but might still have some feelings? Or does he think that without a cortex the lights are completely off, and any such being would have no experience at all, even if it could manage some behaviors? Most animals, especially most of the animals in this book, do not have a cerebral cortex. Do they have experience of a different kind from us, or no experience at all?

Some people do think that without a cortex there can be no experience at all. Perhaps we will be pushed to a view like this in the end, but I doubt it. We need to continually avoid falling into the habit of thinking that all forms of experience must be human-like in various ways. When the word "consciousness" is used for the very broad idea of felt experience, it is easy to go astray. But many people do now use the word "consciousness," or some modification of it ("phenomenal consciousness"), in this very broad manner. I am not going to be fussy about the words, and no terminologies are perfect. In some ways, "sentience" is a good term for the broader concept. We can ask: Which animals are sentient? This is, or might be, different from asking which ones are conscious. But "sentience" is often used for particular *kinds* of experience—for pleasure and pain and related experiences that include a valuation, good or bad. Those experiences are certainly important, and it probably makes sense to think that they can exist without sophisticated kinds of consciousness. But these may not be the only kinds of basic or simple experience. In a later chapter, I will look at the possibility that sensory and evaluative sides of experience are somewhat distinct—registering what is going on might be distinct from evaluating whether it is good or bad. "Sentience" is not usually used for the sensory side of this distinction.

Another term is the unwieldy "subjective experience." The term looks redundant (is there another kind of experience?) and it has no easy adjective, like "conscious" or "sentient." But "subjective experience" points in a good direction, by calling up the idea of a *subject*. In some ways this book is about the evolution of subjectivity—what subjectivity is and how it came to be. Subjects are the home of experience, where experience lives.

I will also talk sometimes just about the mind, as I think that is what we come to understand through this story: the evolution of the mind and how it fits into the world. I'll move between terminologies without laying down the law. Our present understanding is not good enough to insist on one language or another.

The project I am trying to advance can be described in a number of different ways, but it is difficult no matter how we look at it. This project is to show that somehow a universe of processes that are not themselves mental, or conscious, can organize themselves in a way that gives rise to felt experience. Somehow, a part of the world's often-mindless activity folded itself into minds.

Dualism and panpsychism and various other views think that this cannot happen; you can't make a mind—not wholly, anyway—from something else, from entirely non-mental ingredients. Mind must be present in everything, or it has to be added "on top"—not literally on top, but added to a physical system that would be complete, in principle, without it. I think, instead, that you can—or evolution can—build a mind from something else. Given some arrangements of things that are not themselves mental, a mind comes to exist. Minds are evolutionary products, brought into being by the organization of other, non-mental ingredients in nature. That coming into being is the topic of the book.

I said that mind is an evolutionary product and something *built*, but I want to prevent a common mistake from arising right away. A materialist view does not claim that the mind is an *effect* of physical processes in our brains, a consequence or product of them. (Huxley seems to have thought that.) The idea, instead, is that experiences and other mental goings-on are biological, and hence physical, processes of a certain kind. Our minds are arrangements and activities in matter and energy. Those arrangements are evolutionary products; they are slowly brought into being. But those arrangements, once they exist, are not *causes* of minds; they *are* minds. Brain processes are not causes of thoughts and experiences; they *are* thoughts and experiences.

That is the biological materialist project as I see it—showing that such a position makes sense, and is, most likely, how things actually are. The aim of this book is to work as far as I can down this path. I don't think a solution to the problem will be revealed in a single stroke of the pen, in a move that pulls a rabbit from a hat. It will be more cumulative. As this book moves along, I will develop a positive view, a sketch of a solution that combines roughly three elements in a picture that I think makes sense. But not every question will be answered, and many puzzles will remain. The way I think things will go is vividly expressed in a passage that through years of drafts I had as an epigraph to this book. The passage is by Alexander Grothendieck, a mathematician.

The sea advances insensibly and in silence, nothing seems to happen and nothing is disturbed. . . . But it finally surrounds the stubborn substance, which little by little becomes a peninsula, then an island, then an islet, which itself becomes submerged, as if dissolved by the ocean stretching away as far as the eye can see.

Grothendieck worked on very abstract problems—abstract even by the standards of pure mathematics. The quote describes his approach to problems in his field. A puzzle in front of us seems to resist the usual methods. What we should do in response is build knowledge *around* it, expecting that as we do this, the puzzle will transform and disappear. The situation becomes reshaped and eventually comprehensible. The image he used for this process is the submerging of an object, a mass, in water.

I have had that image in my mind for a long while now. I don't think, as some philosophers do, that the puzzles in this area are mere illusions that we can overcome if we just talk a bit differently. New things have to be learned. But as they are learned, the problem itself changes shape and fades.

Grothendieck's image seemed so apt that I once used it to head the book. But the image has new connotations now, at a time when the melting polar ice of a rapidly warming Earth is leading to the loss of precious Pacific islands. Given these new associations, it seemed wrong to begin the book that way. But Grothendieck's metaphor does still guide my thinking, and the perspective expressed there guides how the book will work. *Metazoa* approaches the puzzles of mind and body by exploring the nature of life, the history of animals, and the different ways of being an animal that surround us now. By exploring animal life, we build around the problem and see it transform and subside.

This book is a continuation of a project that began in another, called *Other Minds*. That book was an exploration of evolution and the mind guided by a particular group of animals: cephalopods, the group that includes the octopus. *Other Minds* began with encounters with these animals in the water, scuba diving and snorkeling. Encountering them there, in their protean, color-warping complexity, led to an attempt to understand what might be going on inside them. That led, in turn, to a tracing of their evolutionary path, a path taking us to a pivotal event in the history of animals, an ancient fork in the genealogical tree. That fork, over half a billion years ago, led on one branch to the octopus (among others), and on the other branch to us.

Some ideas about minds, bodies, and experience were sketched in that book, guided by the animals I was following. Here, those ideas are developed and augmented. That development comes as a result of a closer look at the philosophical side, an exploration of further branches of the tree, and watery hours spent with more of our animal relatives. Whereas in *Other Minds* I kept coming back to octopuses, my aim in this book is to move along with many kinds of animals, both closer and further from us on the evolutionary tree. For some of those animals I, too, was a being they could observe and encounter; for others, a presence in less than a dream. Toward the end of the book, we begin to approach nearer relatives, with bodies and minds more like our own. But the historical story is weighted to the earlier evolutionary stages, and its goal is to make sense of how experience came to exist on Earth at all, first in its waters, later on land.

That, then, is the path of the book. We walk—crawl, grow, swim—through the story of animal life from its beginnings, guided by a collection of present-day creatures. We learn from each animal—from its body, how it senses and acts, how it engages with the world. With their aid, we try to discern not just the history, but the different forms of subjectivity around us now. My goal is not encyclopedic, trying to cover every variety of animal. I concentrate on those that mark transitions in the evolution of the mind, especially the stages by which it came to be. Most of these are marine animals, living in the sea. Let's walk down the steps.

2

THE GLASS SPONGE

Towers

A sponge garden often begins just below the layers of water that sunlight penetrates well, especially in places where currents flow. There, as light falls away, you may find a landscape of motionless animal bodies. They have the appearence of cups, bulbs, grails, or branching trees. Sometimes they look like hands in thick mittens, as if something huge beneath the seafloor was trying to reach upward with soft, half-made limbs.

While in this shallow zone, look out and imagine a sea much colder, the scene now a blackness with a thin fall of particles from above. On the ocean floor, 3,000 feet below the surface, a pale tower, about a foot tall and cylindrical, sits in a cluster of other towers, each held fast at the bottom and a little broader, partly open, at the top. Within its soft exterior is a lattice of tiny hard parts. The smallest of these are stars, hooks, or slender crosses, with angles skewed to knit into that tower-like form. The towers are held to the seafloor by delicate anchors. The anchors and crosses are made of silicon dioxide, the main constituent of glass.

A sponge, on a temperate reef or a deep ocean moonscape, looks dead and inert, but if you look closely, it is not. It is a silent pump, drawing water through itself. As it does so, it senses and responds. The deep-sea tower, the glass sponge, has a body that also conducts light and charge, like an electric light bulb ("think!") at the bottom of the sea.

Cell and Storm

The background to the evolution of the mind is life itself—not everything about life, not DNA and its workings, but other features. The start is the cell.

Early life, before animals and plants, was single-celled. Animals and plants are huge collaborations of cells. Before those collaborations arose, cells were probably not entirely solitary, but often lived in colonies and clumps. Still, a cell then was a tiny self of its own.

Cells are bounded, with an inside and outside. The border is a membrane; it partially seals the cell but has channels and ports embedded in it. A continual to-and-fro runs across that border, and inside is a frenzy of activity.

A cell is composed of matter, a collection of molecules. I don't know exactly what comes into your head when I say "matter," but the word often brings to mind an inert, ponderous way of being, with weighty objects needing to be pushed into motion. That picture of matter is guided by how things work on dry land and at the scale of midsize objects like tables and chairs. When we think of the material of a cell, we need to think differently.

Inside a cell, events occur on the *nanoscale*, the scale at which objects are measured in millionths of a millimeter, and the medium in which things happen is one of water. Matter in this environment behaves differently from anything in our midsize, dry-land world. At this scale, activity arises spontaneously, without having to be made to happen. In a phrase due to the biophysicist Peter Hoffmann, within any cell is a "molecular storm," a ceaseless turmoil of collisions, attractions, and repulsions.

If we imagine a cell full of intricate apparatus, parts with jobs to do, these devices are continually being bombarded by water molecules. An object in a cell has a fast-moving water molecule collide with it about every ten trillionth of a second. That's not a typo; the scale of events in a cell is almost impossible to think about in an intuitive way. Those collisions are not trivial; each has a force that dwarfs the forces those devices can themselves exert. What the apparatus inside a cell can do is nudge events in one direction rather than another, bringing some coherence to the storm.

The medium of water is important in maintaining this storm. Many objects at this spatial scale would stick together and seize up in a clump if they were out on dry land, but do not seize up in water; instead they're kept in motion, and this makes the cell a realm of self-generated activity. We often think of "matter" as inactive and inert, I said. The problem cells have to deal with, though, is not getting things to happen, but creating order, instituting some rhyme and reason in the spontaneous flow of events. Matter, in these circumstances, does not sit there doing nothing, but risks doing too much; the problem is getting organization out of chaos.

Nearly all the associations we habitually bring to bear when we think about matter are misleading when we consider life and how it might arise. If life had to evolve on dry land out of table-and-chair-sized ingredients, it could not happen. But life did not have to do this; it evolved in water—perhaps in thin films of water on a surface, but in water—through the emergence of order in a molecular storm.

The origin of life occurred fairly early in the history of Earth, perhaps around 3.8 billion years ago, on a planet now about 4.5 billion years old. The first life may not have been cell-shaped, but there must have been some initial way for a special set of chemical processes to be contained, marked off, and prevented from diffusing away. Then at some stage there were cells, presumably leaky and tenuous at first, but eventually arriving at something like bacteria, cells that consistently maintain their organization and reproduce.

As cells acquired the power to keep themselves going—transforming materials, imparting order, bringing method to madness—a central achievement was gaining control over charge.

The Taming of Charge

The taming of electric charge was a pivotal event in recent human history. In the nineteenth century, electricity went from being a mysterious, often dangerous force—encountered most directly in lightning—to an element in technologies that soon formed the modern world. If you are reading this book under electric light or on a computer, the act of reading is electrically sustained. This modernizing electrical advance was the second of two. Charge was also tamed billions of years before, during early stages of life's evolution. In cells and organisms, electricity is the means by which much of what happens is done. It is the basis for brain activity—our brains are electrical systems—and also a great deal else.

What *is* electricity? Even many physicists find this question elusive. Electric charge is a basic feature of matter. Charge can be positive or negative. Objects with the same charge (positive and positive, for example) repel, and those with unlike charges (positive and negative) attract. The stuff of ordinary objects contains both. Any atom is a combination of even smaller particles, some that are positive (protons), others that are negative (electrons), and in most cases, other particles (neutrons) with no charge. Usually, an atom will contain the same number of electrons as it has protons, so the atom itself will have no net charge, as the positive and negative charges within it are exactly balanced.

The electrical tendency to attract and repel is strong. Here is the inimitable Richard Feynman, in his *Lectures on Physics*.

> Matter is a mixture of positive protons and negative electrons which are attracting and repelling with this great force. So perfect is the balance, however, that when you stand near someone else you don't feel any force at all. If there were even a little bit of unbalance you would know it. If you were standing at arm's length from someone and each of you had *one percent* more electrons than protons, the repelling force would be incredible. How great? Enough to lift the Empire State Building? No! To lift Mount Everest? No! The repulsion would be enough to lift a "weight" equal to that of the entire earth!

In the mix of charged parts that comprises ordinary matter, electrons, the negative particles, are on the outside of atoms, while protons (along with neutrons) are on the inside. Electrons on the outside can sometimes be gained or lost, resulting in an *ion*. An ion is an atom (or sometimes a molecule that combines a few atoms) that has unbalanced its charged parts through such a loss or gain, and hence has an overall charge of its own. When many chemicals dissolve in water, they produce ions that then drift around. *Salt* water is water with dissolved ions. Any droplet of seawater will contain countless ions, interacting with each other and with the water molecules, attracting and repelling.

An electric *current* is a movement of charged particles, either positive or negative. In a metal wire, a current takes the form of a movement of electrons, with the rest of each atom making up the wire remaining in place. The electric currents used in technology (lights, motors, computers) mostly work in this way. But a current can also be a movement of whole ions. If some positive or negative ions in water, for example, can be induced to move in a consistent direction, that is an electric current. It does not make a current flow; it *is* one. Any container of salt water can contain such a current, if you can somehow get an overall pattern of movement of ions of the right kind to occur. In living systems, unlike human inventions, most currents take this form.

Charge is not life-like or mental in itself. It produces much of what happens in the inanimate world as well as the animate. But living activity runs on charge, especially by the corralling, pumping, herding, and unleashing of ions.

A cell's membrane keeps many things either outside or inside, but it contains channels that selectively let some material through. Many of these are *ion channels*. Sometimes a channel will passively allow ions to move from one side to the other, perhaps under specific circumstances; in other cases, the cell pumps the ions across the membrane.

Ion channels are shared, with variations, across all kinds of cellular life, including bacteria. The reasons for bacteria to build elaborate ports and passageways for ions are often not entirely clear. Channels may have arisen initially just to enable cells to adjust their overall charge in relation to the outside—tuning as well as taming their charge. Whenever there is traffic across a living system's boundaries, though, it tends to take on further roles. A flow of ions can function as a minimal form of sensing, for example: suppose contact with a particular external chemical opens a channel and lets in ions. Those charged particles can set new events in the cell in motion.

The next consequence of these ion flows is related to that to-and-fro traffic, but is a larger, more wholesale change to the cell. This next step is *excitability*. Channels control the flow of charged particles, and these channels can themselves be controlled: they can be opened or closed. This can happen through chemistry, or physical impact, but it can also involve charge itself. Voltage-gated ion channels are channels that open as a response to electrical events that they, the channels, are exposed to. This makes possible a chain reaction; a flow of current creates a greater flow of current, one that spreads over the cell membrane.

This might not seem much of a step, and it has less of an obvious ring of usefulness than the arrangement I described above, where the flow of ions is sensitive to chemicals the cell encounters in its travels. But voltage-gated ion channels are the basis for another innovation, the *action potential*. This is a moving chain reaction of changes to the membrane of a cell, especially in our brains. Positive ions flow into the cell at one point, affecting ion channels at adjacent points, which open and allow more ions to come in, and so on. A wave of electrical disruption travels along the membrane like a pulse. An action potential is the zap-like event that is described as a brain cell "firing." That zap happens by means of voltage-gated ion channels.

In a voltage-gated ion channel, a controller of current is affected by charges it is exposed to; the flow of current is electrically controlled. This is the principle of a transistor. At the start of this section I mentioned the nineteenth-century advances that brought electricity into the realm of human technology. Another such advance occurred in the twentieth century, with the invention of the transistor. The silicon chips in computers and smartphones are collections of tiny electrical switches of this kind. The transistor was invented around 1947 at Bell Laboratories in the United States—or invented once then, anyway. The first Bell Lab transistor was an inch or so in size, and it has been continually refined and shrunk since then. The same device was invented billions of years ago in the evolution of bacteria.

If bacteria invented transistors, what were they doing with them? Why did *they* need to control electricity with electricity? As far as I can tell, no answer to this question is widely agreed on. Bacteria might have been using them as part of the electrochemical upkeep of the cell. They might have been used in the control of swimming. Channels that sense external chemicals may be incidentally sensitive to charge, and bacteria that form colonies in "biofilms" signal from cell to cell using ions. But bacteria don't have action potentials—the zap-like chain reactions in our brains—and the situation does seem quite odd to me. Several billion years ago, nature invented the fundamental hardware device in computer technology—a complicated and costly device, too—and did so in bacteria, but bacteria do not seem to have been doing much computing with it.

Regardless of why it arose, the voltage-gated ion channel was a landmark in the taming of charge. These channels do not have a single obvious use, I said above. In a sense, neither does a transistor, and in both cases, that is part of their importance. A transistor is a general means for control, a device for making events *here* affect events *there* in a reliable, rapid way. The events controlled can be multifarious—whatever might be useful. When they enable action potentials, voltage-gated ion channels also make it possible for a cell's activity to have a "digital" quality; a neuron either fires or not, yes or no. Not all animals have neurons with these zap-like firings, and nervous systems can work with milder kinds of excitability, but this digital feature is certainly a useful one. It is remarkable that this control device was invented so far back, when most of the uses it has now were not even glints in evolution's eye.

In the days of ubiquitous computers and AI, it is natural, almost inevitable, to ask about the relationships between living systems and these artifacts. Do organisms and computers do essentially the same thing with different materials? Similarities between the two do arise, often unexpectedly, but it's also important to recognize dissimilarities. One difference is that much of what a cell does, its main business, is something a computer never has to do. A great deal of the activity in a cell is concerned with maintaining itself, keeping energy coming in, keeping a pattern of activity going despite decay and turnover in materials. Within living systems, the activities that look like the things computers also do—electrical switching and "information processing"—are always embedded within a sea, a mini-ecology, of other chemical processes. In cells, everything that happens takes place in a liquid medium, subject to the vicissitudes of the molecular storm and all the chemical digressions that living systems engage in. When we build a computer, we build something whose operation is more regular and uniform; we build something that will be distracted as little as possible by the undirected ruminations of its chemistry.

This relates to a broader point. Often in these early chapters, I'm trying to describe a tangle of parts and processes inside cells and simple organisms. A natural word to use at many stages is "machinery"—we're looking at the machinery of sensing, the machinery of excitability. I type the word "machinery" and am never sure whether to delete it. In a broad sense of the term, yes, voltage-dependent ion channels are bits of machinery, and so are nerves and brains. To deny this is to suggest a move toward dualistic (soul + body) or vitalistic ("life force") views. So, I say to myself, don't delete the word. But the contrasts between machines and living systems are also important. In cells, the processes of life involve imparting order upon a molecular storm and the imperfect herding of ions. This is nothing like what goes on in any machine we've built. We generally build machines to be predictable and restricted in their activities, even if we might then use them to simulate more chaotic goings-on. To describe the intricate materials in cells as "machinery" is right in some ways and wrong in others.

There is one more thing I want to emphasize in this inventory of features of life that were in place before animals. This one has been touched on a few times above, but I now want to put it at center stage for a moment. That feature is *traffic*, a to-and-fro between living systems and their surroundings. This traffic includes the flow of ions described above, also the taking in of raw materials and elimination of waste. Cells are bounded, but they are not closed to the world. I am emphasizing here the *windowed* character of cellular life.

This traffic has a metabolic side—a side that involves getting energy and using it to stay alive—and also an informational side. Some incoming influences are important in their own right (as food, for example), while others are important for what they predict and portend, for what they indicate about something else. The metabolic side of this to-and-fro is unavoidable if life is to continue. Living activity itself is a pattern that exists embedded in an energetic flow, one that begins and ends outside of the organism. My colleague Maureen O'Malley expressed this well; combining some chemical jargon with an image from a different source, she said that being alive requires learning how to exist "on a redox rollercoaster, perpetually giving and receiving." (A redox reaction is one involving a transfer of electrons between two molecules.) A consequence of this, and part of what O'Malley wanted to emphasize, is that living systems are inherently sensitive to changes and events outside. They don't have the option of being windowless, but are open to the world out of energetic necessity. Once open to their world in this way, they will be affected by what goes on. And once they are affected by those goings-on, evolution will tend to put this sensitivity to use—the organism will often find a way to react to events in a manner that furthers its projects, simple as these might be. All known cellular life, including tiny bacteria, has some ability to sense the world and respond to it. Sensing, in at least the most basic forms, is ancient and everywhere.

Metazoa

Those ideas complete one of the two themes of this chapter. Living cells are physical objects, but unlike any other object we are familiar with. They build membranes to contain and shape storms of activity. They are bounded, but forever dependent on traffic across those boundaries. Self-defining, self-maintaining, cells are *selves*. The next step in the story takes us to a new kind of unit, a new kind of self: animals.

When we think of animals, we usually first think of animals like us—other mammals, dogs and cats, perhaps birds. But animals extend much further than this. Animals—the Metazoa—form one large branch of the total tree of life, the genealogical network that links all life on Earth. The term "Metazoa" was introduced in the late nineteenth century by Ernst Haeckel, the German biologist of Chapter 1. He contrasted Metazoa, multi-celled animals, with Protozoa, single-celled animals (with "zoa" as in "zoo" and "zoology"). The Greek prefix "meta" originally had meanings like *after* and *beside*, then took on the connotation of *higher*, and now often means *about*—looking down on. Haeckel probably had in mind some mix of higher and later. But proto-zoa are no longer considered animals at all, so the "zoa" part of their name becomes misleading. The animals are now just the Metazoa.

Animals are made up of many cells living as a unit; beyond that, they live in a huge variety of ways. They include corals as well as giraffes, wasps that are smaller than some single cells as well as whales at fifty tons. Some look almost entirely like plants. In biology now, the word "animal" refers to any organism found on a particular branch of the genealogical tree, regardless of how it lives or what it looks like. A coral is as much an animal as a wolf is. This is not the only meaningful way the term "animal" might be used, but it is unambiguous and clear, unlike various other uses.

Animals do not form a scale from "lower" to "higher," though the habit of talking about them in this way seems hard to break. On the genealogical tree, some animals are lower in the sense of earlier, but insects that are alive now are not lower than us; everything alive now is at the top of the tree. So there's no sense in talk of an evolutionary "scale" or "ladder"; animal life has a different shape from that. Some animals are more complicated than others, in various ways (more parts, wider range of behaviors, more complicated life cycle . . .), but biology has no room for an overall scale from lower to higher, of the kind that seemed natural before Darwin.

The genealogical network that animals are part of—the "tree of life"—is not always tree-shaped; in many places it is more tangled than that. For simplicity, I will keep referring to it as a tree. This tree links all known life on Earth by relations of ancestry and descent. It is old now, but continues to grow. This growth occurs through evolutionary processes operating over huge spans of time. Populations or species occasionally split into two. Each side then evolves on its own path and acquires its own peculiarities. Extinction is always likely, but any segment—a new species—that doesn't go extinct may later split again. From a single initial fork, we then have several branches, each with a collection of species rather than just one.

Many years ago, when the tree was younger and smaller, an outgrowth protruded: a new twig. The twig survived, branched repeatedly, and became particularly far-flung and diverse. The organisms on that part of the genealogical tree are the animals. Evolution is open-ended, and there is no telling where future branches might extend, both within and outside the animal part of the tree. But though animals have lived in a great variety of ways, there is a general *style* of living seen in animals, a way of life invented on the animal branch of the tree.

Animals arose from a particular kind of single-celled organism, larger and more internally complex than bacteria. These cells, *eukaryotes*, have special devices for handling energy—mitochondria—and an elaborate internal skeleton (the cytoskeleton). This is an internal network of filaments and tubes that can move in relation to each other, enabling the cell to control its shape and motion.

Well before animals arose, the cytoskeleton had initiated a new regime of mobility in single-celled organisms, including active hunting. This apparatus made possible a shift from an existence based primarily on chemical processing, as seen in bacteria, to one based partially on behavior: motion and manipulation. All of these sound like animal characteristics, but we are still talking about single-celled organisms—protists. Some of them have grown large. Members of the genus *Chaos*, for example, hunt not only bacteria but, in some cases, small invertebrate animals.

Plants are another branch of the genealogical tree, another long-term multicellular experiment, and they too are collections of eukaryotic cells. So are fungi. A recurring theme in evolution is the formation of new and larger units by the collaboration of smaller ones. The eukaryotic cell itself came to exist in this way, through the engulfing of one simpler cell by another. The engulfed cell gave rise to the mitochondria that eukaryotes use as powerhouses.

In the events that gave rise, separately, to animals and plants, another kind of coming-together occurred, this one not an engulfing but a juxtaposition. Suppose a single cell divides, and rather than going their own way, the two daughter cells instead stay stuck together, as the result of a mutation that affects their chemistry. When those cells divide, their daughters will stick together as well. The initial result is just a bigger living object. This object cannot act as a whole, and has no obvious way to reproduce, as opposed to getting bigger. But this is a step toward a new kind of life.

Multi-celled beings of this kind have evolved from one-celled forms repeatedly. On the animal line, this might have happened about 800 million years ago (with a good 100 million years of uncertainty). There are no fossils of the earliest forms, but we can picture the first stages: a ball of cells in the sea, formed by a succession of cells refusing to separate from their sisters.

Where to from there? One tradition of speculation imagines a cup, or a hollow sphere with an opening, as a likely next stage. The ball of cells folds in on itself and becomes hollow. This possibility was also first sketched by Ernst Haeckel.

One reason the cup hypothesis is tempting is that this form is seen in the early stages of individual development—development from egg to adult—in a wide range of animals. The hollow form is a *gastrula*. It is a mistake to think that something seen early in individual development must also have been there early in evolution (as Haeckel did suppose), but the cup form seems so ancient and widespread that it might be a clue. Haeckel christened this hypothetical animal the "Gastraea."

The *Bathybius* affair of Chapter 1 was not Haeckel's finest hour; the Gastraea was better. It is still a live possibility for a very early animal form. The open sphere might have been the beginnings of a gut; the first animal might have come to exist by forming around its stomach. In that enclosed environment, it could trap food and release digestive enzymes, without having them drift away.

A human gut holds our food. In addition, our guts contain countless living bacteria, from which we benefit greatly as long as things stay in balance. This kind of collaboration is extremely common in animals. It may have also been part of the early stages in animal evolution. That idea was not part of Haeckel's original version of the view, nor part of most thinking since then. It is a newer idea, informed by the realization that normal animal bodies are homes for large colonies of bacteria that help them process food as well as playing other roles. The recognition of ubiquitous tight associations between our bodies and accompanying microbes has been a significant shift in how biologists think of animal life, and perhaps these associations go very far back. Remember also those engulfing events in the history of cells, the events that produced mitochondria, and also chloroplasts within plants. In those meetings, a metabolic ally was brought inside a cell—or first brought in and then tamed. Here, in comparison, we build a home for collaborating microorganisms without having them enter our cell bodies; instead we build a pen for them. A diverse digestive ecology might be at the beginning of animal life.

This open-sphere idea, with or without collaborating microbes inside, is like a second iteration of the evolution of cells. In the first case, we had the formation of a boundary, with channels across it, forming a unit that controls chemical reactions. Here, we have many cells and they form a hollow sphere, another object with an inside and an outside. Individual cells are each now pieces of the sphere, and they control traffic in and out of this larger unit.

From there—or somewhere—early animal bodies gained more shape. For those trying to work out the next steps, the fossil record is still frustratingly silent as I write these words. But we do have clues in some present-day animals. These clues are easy to misread; the present-day animals are not preserved ancestors, but distant cousins. They have been through as many years of evolution as we have. But some of them might have stayed in a form that resembles old forms, in some ways, or at least indicates something about them.

The animals that contain these clues are a trio: sponges, comb jellies, and placozoans. They are quite different. A sponge, once it has settled, does not move as an adult. It lives fixed in place like a plant. Some sponges also grow very large. Placozoans, in contrast, are tiny, flat, crawling creatures with little definite shape. You need a microscope to see one clearly. Both sponges and placozoans have no nervous system. Comb jellies, as the name suggests, resemble jellyfish, but may be a considerable evolutionary distance from them. They do have nervous systems, and they swim using cilia, tiny hairs along the side of the body that beat in rhythm. So of these clues, one is a piece of motionless undersea furniture, one crawls nervelessly and microscopically, and the third is transparent and swims.

Why are these, among animals, the clues about early forms? First, they are simple animals in various ways. They have few parts and not many kinds of cells. Second, they are genetically very far from us. On the genealogical tree, they are on lines that branched off from our line very early.

It is worth pausing to think about that combination of features—being simple and being far from us. There is no general reason why these should be associated. We might have found, on Earth today, an extremely complicated animal whose evolutionary path turned away from ours very early. All the time we have had to evolve our complex bodies and brains is time that this other animal would have had, too. The best partial example of this combination—complicated, far from us—is the octopus, waiting in a later chapter of this book. But octopuses are not nearly as far from us as sponges and the other animals we are talking about now.

It has often been tempting to tell a story in which some of our earliest animal ancestors looked like a sponge, then later ancestors looked like a jellyfish, and so on. That sequence is not impossible, but it can't just be read off the evolutionary tree. To do this would be to treat a set of cousins as if they were grandparents—or to treat distant cousins as if they looked more like grandparents than other cousins do. Once we put the point in terms of cousins and grandparents, it is clear that this reasoning would not make sense. However, there can be other reasons that some particular distant cousins might contain clues.

We, in our bodies, possess various evolutionary inventions (brains, hearts, backbones, and so on) that had to come into existence somehow. Sponges and jellyfish live without those inventions, though they do share ancestors with us. So these animals show, first of all, what you might be like if you had to live without them. Further, these animals are not part of evolutionary lines that had these features at some stage but lost them; it is pretty clear they never had them. The inventions they lack include more than mere accessories, too. The left-right body layout that we have is an invention. The intricate folds of tissue that make up our internal organs are an invention. Looking at distant animals who lack these inventions, along with genetic evidence and fossils, we can start to get a sense of what ancestors far below us on the tree might have looked like.

Light Through Glass

Historically, sponges have been seen as the most important of the living clues to very early animals. Sponges have a decent fossil record and are the best known. So without making assumptions about whether the sponges around now look much like any of our ancestors, instead approaching them just as their unusual selves, let's take a closer look.

Sponges extend widely through the seas, from fingers and trees in temperate waters and huge funnels on tropical reefs to the freezing deep-sea towers that head this chapter. Some encrust themselves on other organisms rather than growing a shape of their own. Their modus vivendi in many cases is that water is drawn into their lower parts, passed upward through the body, and exits at the top. Food, mainly bacteria, is taken from the water as it passes through. A few sponges have a more ambitious diet; the deep sea has some predatory sponges that trap and consume small animals.

A sponge has a body, but one with a big difference from bodies like ours. Most of its cells are in direct contact with the water moving through it. The sponge body is a maze of fine passageways, lined with microbial partners, and the environment suffuses that body.

A sponge has no brain or other nervous system. The larvae (immature forms), which look like tiny fat cigars, can swim, and they have some sensing structures that resemble parts of a nervous system. These sensing mechanisms are facing out to the world, not toward other cells in the same body. The larva settles and the adult grows up in place. Though a sponge has no nervous system, it is not inert. Inside each cell is the storm described earlier in this chapter. A sponge as a whole seems a good deal quieter, but it has something of an active side.

Water moving through the sponge's body is actively pumped by cells with little tails (flagella). This pumping can be modified or stopped, especially if the water is dirty and the sponge's channels risk becoming clogged. For a collection of cells with no nervous system, this is not easily achieved. There appear to be specialized sensory cells along the chimneys through which water passes, and these cells signal to others. Given what cells are, it is a significant task for one to influence another. The usual way this is done is with the release of small molecules from one cell, taken up by others. The result is to contract, squeeze down, the canals. This process is slow, but there is little need for it to be fast. In some cases, the sponge expands a little first and then contracts, in a drowsy and invisible "sneeze."

All this is a reminder of both the opportunites and difficulties of multicellular life. A cell within a sponge is at little risk of being eaten by a larger cell, as it would be if it floated freely in the water. If the cell was merely locked in place with others, though, it could easily starve. In a sponge, a filigree of channels and chimneys keeps most cells in contact with water. But if something then has to be done, it is hard to achieve coordination, especially coordinated motion. All in all, the result looks a fair bit like a plant. Most sponges are entirely content with that, and have been doing what they do for many more years than us. Still, a few tried something different.

The Hexactinellida, or glass sponges, explore in their bodies this chapter's themes of unity and selfhood in a unique way. A glass sponge is a multicellular organism, like other animals, but as it grows up, most of its cells fuse, losing their boundaries. They don't abandon boundaries with the outside world, but with neighboring cells. The body eventually forms a single connected net, often described as a "three-dimensional cobweb," draped over harder elements that support it.

Those harder elements are made of glass. Variously, according to species, they resemble daggers, stars, or snowflakes.

These are grouped together to form flowers, vine clusters, and—finally, in combination—the skeleton that holds up the tower. (The drawings of these tiny parts below, by Rebecca Gelernter, reproduce engravings made from specimens collected on the nineteenth-century *Challenger* expedition, the voyage that doomed *Bathybius*.)

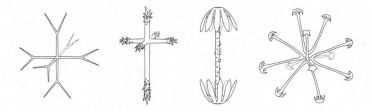

Like other sponges, the Hexactinellida live in rich associations with other kinds of life. The Venus's flower basket is a glass sponge usually found with a pair of small shrimp living inside its body. These shrimp enter the sponge when they are very small, and grow up there, never leaving. Eventually they become too big to slip out through any opening. The shrimp eventually reproduce together. They keep the sponge clean, and benefit in turn from both the protection of the sponge's skeleton and the food in the water the sponge draws through its body.

Glass sponges have no nervous system, but they are not electrically inert, and here the taming of charge takes an unusual form. With that fine living cobweb draped over a skeleton, they, uniquely among sponges, have electrical signaling and a kind of "action potential." A glass sponge usually moves water through its body in a continual flow. In response to some stimuli, though, such as the plucking of a single glass element from its body, the pumping quickly ceases. This is done by sending an electrical pulse through the body. Electrically, a glass sponge behaves like a single enormous cell; a pulse can run uninterrupted over the whole. A glass sponge achieves coordination not by coordinating signals between cells, but by being, in large part, post-cellular. It is a product of the animal evolutionary path, but one that has partially abandoned the multicellular form of life, living by means of a different kind of unity.

I have been talking about charge, communication, and co-ordination in these creatures. But this is also an animal made largely of glass—not the charge-conducting cobweb, but the skeleton beneath. A notable feature of glass, of course, is how it handles light. Parts of the skeleton of some glass sponges resemble fiber-optic cable, along which light is conducted and filtered.

Is the sponge doing anything biologically significant with light, or is this conduction just a consequence of employing glass as a building material? Is light *used*, or just channeled around by happenstance? A wide and rather wonderful range of possibilities has been raised and discussed for different sponge species. Light, which except in the shallowest cases will derive mostly from bioluminescence of one kind or another, might be a further means for communication within the animal. It might also be feeding some of the microorganisms within: tiny diatoms and other creatures that live on light huddle deep inside some sponges, perhaps so deep that they could not get enough light to stay alive if the sponge did not channel rays to them. Light in these cases is even ducted a little way into the seafloor itself. The Venus's flower basket emits light into the surrounding water, perhaps as a faint lamp in the ocean that attracts the shrimp who come to live inside. These possibilities are unresolved, and some biologists think the light in and around the sponges will be too weak to do much. Either by design or by accident, a glass sponge is a collector and curator of biological light.

THE ASCENT OF
SOFT CORAL

Ascent

In a bay north of Sydney, Australia, a short way from the stairs we walked down in Chapter 1, is a sandy underwater plain. The bay is formed where a river winding down from eucalyptus forests inland reaches the Pacific.

The underwater plain experiences a strong tidal flow. Water pushes hard up the river from the ocean as the tide rises, and runs back out to sea as it falls. This flow of water attracts a rich range of animals, but also means that the area can only be visited by scuba divers for a couple of hours each day, around the periods when the water comes to rest at the turn of the tide. There is about an hour at each pause. You dive on top of the high, until the water starts to run.

The turn of the tide comes on fast—a tug and then you are moving. Soon it is not possible to swim against the flow. If you stay too long, you can quickly be pulled out to sea.

A few areas on the plain have fields of purple and white soft coral. This coral is soft and wispy, not jagged and mineralized like tropical "hard" corals. The corals form trees that have cauliflower shapes, though a comparison to cauliflower does them no justice at all. From a distance they look like puffs of white and purple cloud; up close one sees delicate veins and filaments, with cowrie shells and crabs living among the branches.

If you approach on a slight current, perhaps riding the end of the incoming tide, it is like coming in on a glider, silently approaching clouds, and finding that the clouds are growing from the earth on stout pale stems. These trees are not single organisms, but colonies of many small animals, coral polyps. From Chapter 2 we know there is an unending frenzy of microscopic activity within them. But the corals seem to sit motionless while more active animals crawl among their branches.

A few years ago, a local diver and researcher, Tom Davis, who had dived in the bay countless times on the still of the high tide, wondered: What do the soft corals do when no one is looking? Through most of any day the tidal flow is too strong for a diver to be down there with them, but he could put cameras down, and take time-lapse footage of what goes on when the water is fast and no humans are around to see.

Diving with his wife, Nicola, he dropped cameras in various sites where the coral is found. When the cameras were retrieved and they watched the results, they found that as the waters sped up with each change of tide, the soft corals slowly rose, by inflating their bodies, until they were up to three times their still-water size. Very likely, they were stretching out to take in the extra food swept past on the tidal flow. The corals subsided when the waters slowed, and sat lower during the hour or so when humans might return.

In Search of the First Animal Actions

Corals are *cnidarians* (again with a silent "c," so "nye-dairians"), the same group of animals that includes jellyfish and anemones. This group diverged from our own evolutionary line at a time still early in the history of animal life. A coral might have last shared an ancestor with you 650 or 700 million years ago. The date is unclear, but it was certainly more recent than the time you last shared an ancestor with a sponge.

A cnidarian body is soft, radial in organization—organized around a disc or cup—and often fringed with tentacles. These can be long streamers or short fingers. Inside the body is muscle, and the electrified threads of a nervous system.

Many cnidarians have a complicated life cycle, proceeding through a series of different bodily forms. These transitions are a bit like metamorphosis, as in caterpillar to butterfly, but not quite the same, as the bodies not only transform but multiply at more than one step; it is as if a caterpillar could make many butterflies, as well as a butterfly many caterpillars. The two adult forms a cnidarian can have are the polyp and the medusa. A polyp is usually fixed to a surface, often shaped like a cup. The medusa is the familiar jellyfish body, swimming in open water with streaming tentacles. Many cnidarians cycle between these two bodies. Corals and anemones only live as polyps.

A little way from the cloud-like trees on the plain, another kind of soft coral lives on a tangled reef. These corals sometimes form shrubs but also cluster in irregular masses. Each polyp looks like a white flower with eight long, finger-like tentacles. Each finger, in turn, has smaller rays extending from it sideways. Those are called *pinnules*. Fingers on fingers. A coral colony is often partly covered by an orange sponge, growing as a blanket and allowing the flower-like parts of the polyps to reach through.

Having eight of the finger-like tentacles, these animals are called "octocorals." A mass of them together forms a forest of tiny hands. As you watch, if you are patient, often you will see a slow opening of a polyp like a reach, or a closing of the fingers, a clench.

Occasionally a single tentacle will curl and close while others stay extended. At other times the whole hand will close. On a larger scale, you might come across zones or outcrops where every hand is closed, while a neighboring zone has most hands open. They look like they are reaching and grasping *for* things, but for a while it was unclear what, if anything, they were catching. John Lewis, a Canadian biologist, looked at thirty octocoral species and found that some were indeed catching not only plankton, but tiny invertebrate animals. This talk of reaching and seizing suggests a momentary action that a person might make, but in general, the process occurs in a graceful slow motion: faster than a plant, but slower than the scale of familiar, busy animal behaviors. In these unfurlings and clenches, there are hints, indicators, echoes of the beginnings and simplest forms of animal action.

Why might this be? First, cnidarians are old animal forms, with body designs that are very possibly part of our past. It is not clear that any modern-day cnidarian—anemone, coral, jellyfish—closely resembles an ancestor of ours, but that radial layout probably does resemble the layout of some of our predecessors.

Second, they can act. Action itself was not invented by cnidarians. Many single-celled organisms can swim, with propeller-like flagella or hair-like cilia. Some can envelop prey and transform their shape. Stirrings into motion can be seen across all the candidates for early animal forms. In the previous chapter we saw the controlled pumping of water through sponges. This is akin to an action and it might be very old.

Evolution is full of gray areas and partial cases—the *first* of something is rarely clear. Evolution also sometimes sees the rediscovery of something old at a new level or scale. In single-celled life, action exists; there is swimming, seizing, and engulfing. These actions may have been an important spur to the evolution of multicellularity itself. The pre-animal world was a world of single-celled predators and prey, and one option for those wanting to avoid ending up as prey is to become too big to comfortably engulf. Then when cells have come together to make animals, action has to be reinvented at the larger scale. New kinds of coordination are needed. The fitful sponges are on the edge of this rediscovery, a partial case. In cnidarians, action exists again unambiguously, with motion and rearrangements on the grand scale of the animal body.

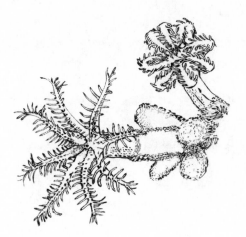

Those reaches and clenchings are not all that cnidarians can do. Another important capacity, an old action of a different kind, is the firing of their stinging cells, nematocytes. All or just about all cnidarians have stinging cells. Some of the stings, especially on anemones, are so feeble they are hardly noticed by a person. Others, on box jellyfish, can kill us outright. The stingers vary but are similar enough in form that they probably came from a single invention early in the cnidarian part of the tree, then passed down many branches.

In the dramatic and sometimes dangerous cases, a harpoon is coiled inside a cell. The cells with harpoons are surrounded by sensing cells and other controllers in a "battery" (this seems to be an artillery metaphor, and an apt one). The harpoon, when triggered, reaches extraordinary accelerations and high speeds over a tiny distance. In a harpoon release, the behavior itself— the motion executed—is performed by a single cell. That cell has helpers around it, sensors and the like, but no coordination between cells is needed for the act itself. Compare this to the soft coral's reach. This is not the action of a cell, but a combination of countless contractions by different cells, motions that have to occur together in a concerted way. What I am highlighting here, marking as a special invention, is action that involves coordination across vast scales from a cell's point of view. The origin of *this* is what we find echoed in the soft coral's reach.

Even if these cnidarian behaviors are echoes of early animal actions, why do I choose this one? What about the coordinated swimming of a medusa? The medusa stage is often seen as a later addition to the cnidarian way of life, with the polyp stage evolving earlier. But a more important point is that if you look at a jellyfish swimming and a soft coral reaching, these are in a sense the same motions. Whether swimming or grasping, in a bell or a cup, both motions involve contraction around a radial form. A polyp and medusa can look very different, but in basic layout, a medusa is a polyp upside down. In the medusa, a radial contraction results in a swimming motion; in a polyp, the animal is stationary and the result is a grasp.

If we are looking for the "first actions," another question is why we should focus on movement, as opposed to the other main thing living organisms do, which is making chemicals. Moving parts of your body and brewing chemicals are both ways of achieving effects that help you make your way. That is true, but the advent of controlled motion on the scale of the animal body was still a landmark. Though action was not invented from scratch by cnidarians, we see in these animals action of a different kind and on a different scale. The bodies that enable these actions were a new kind of object in the world, and a new factor in making things happen.

The Animal Path

The outgrowth of the tree of life that became the animals fairly quickly accumulated a suite of innovations. Perhaps the most important was the nervous system.

Of the animals we have looked at so far, nervous systems are present in cnidarians and comb jellies, but not in sponges or placozoans. Nervous systems evolved early, perhaps just once, perhaps a couple of times. The basis for what nervous systems do is a pair of features that were present long before animals. These are the electrical "excitability" of cells—the capacity for a rapid change in electrical properties, discussed in Chapter 2—and chemical signaling between one cell and another. Nervous systems put those two ancient capacities together. When a cell excites—a sudden shift in its electrical properties—this event is usually confined to that one cell. It is restricted by the boundaries that mark off the cell as a unit. But one thing such a spasm can do is trigger the release of a chemical at the cell's boundary, a chemical that can be picked up by another nearby cell. This in turn may make the second cell more (or less) likely to go through its own electrical changes. Chemical signaling together with excitability are central to how a nervous system works.

Nervous systems are full of cells that specialize in these sorts of interactions. These cells take a tree-like shape, featuring fine projections that bring one cell into chemical contact with a selected group of others. Nervous systems are usually said to be found only in animals (and most, not all animals), though cells with a combination of excitability and chemical signaling are found in other organisms as well. What makes nervous systems in the full animal sense special is those cells with the branching shape: neurons. These are found nowhere outside of animals. Having such cells changes the way influence works within a body. They enable fast and targeted interactions, in contrast to more diffuse patterns of influence in which a cell sends chemicals out to whoever is listening. A nervous system ties the body together in new ways. Lars Chittka, a biologist who works on bees, has an effective way of illustrating their power. A bee, he says, has a brain that is a cubic millimeter in size. That seems tiny. But, he adds, a single bee neuron can have the complexity of a full-grown oak tree, with respect to its branching. Each neuron can connect to ten thousand others.

The nervous system is a reworking of capacities seen over much of life, but in animals these capacities have been extended and made vastly more powerful. A useful reminder of how much these systems do for us is seen in the fact that "nerve toxins" are the main kind of fast-acting toxin, both in animals like snakes and in human misdeeds—notorious weapons like sarin, VX, and Novichok are nerve toxins. On first hearing about nerve toxins when I was young, I thought: It prevents you from feeling anything? You go numb? You can't think? But nerve toxins shut down more than that. Death occurs often by asphyxiation or cardiac arrest. Our vulnerability to these chemicals—which are not intrinsically destructive, which do not lay waste to tissue, but instead interfere with tiny sites of interaction between cells—is indicative of how a nervous system ties the animal body together. A way to kill that body is by going after the messengers, hence going after the coordination of parts.

Another feature closely associated with the nervous system in evolutionary terms is muscle. A cnidarian's actions, unlike the fainter motions of a sponge, are based in muscle. The previous chapter noted the invention of the cytoskeleton, an internal skeleton of mobile filaments inside some single-celled organisms. In animals, the coordination of those internal skeletons across many connected cells generates the innovation of muscle. It enables the coordinated contraction, and relaxation, of massive sheets of cells.

Animals can achieve some action without muscle. In comb jellies, the body is lined with bands of the hair-like cilia seen also in many single-celled organisms. The cilia are lined up vertically in what looks like a comb (hence the animal's common name). Their motions enable the comb jelly, like its single-celled counterparts, to swim. (Comb jellies do also have muscle, which they use for steering.) Cilia are employed for small-scale motions in many other animals. But all the more extensive motions—the octocoral reach, the medusa's swim, and other motions to come—are achieved through muscle.

In discussing the innovations that enabled animals to eventually play such an unusual role on Earth, I have been emphasizing new capacities on the side of *action*. Another animal feature, one I've not been talking much about in this chapter, is *sensing*. Sensing is always present, not just in animals but in all known cellular life. But some of the clues we have suggest that in these early stages in animal evolution, the conspicuous, unprecedented innovation was the creation of action on a new scale. This was the transforming factor.

Cnidarians today have a range of senses, and the same is true of likely ancestral animals at all historical stages. But cnidarians are "skinnier," so to speak, on the sensory side than they are on the side of action. No eyes at all are found in corals or anemones, and only rather rudimentary eyes in most other cnidarians. (The big exception, box jellyfish, are thought to be a later product.) A polyp's reach for food, a colony's ascent and descent, and the firing of stinging cells are all tied to stimuli of various kinds, and cnidarians probably also invented a sense of balance, or a gravity sense. A medusa orients itself in the water by means of organs containing small crystals, called *statocysts*. The crystals are heavier than water, and when they move in response to a shift in the animal's position, those motions can be detected. There may be other subtle forms of sensing, too, but in cnidarians, sensing is not what might be called their particular richness, their step into the unknown, the thing they do differently. Instead, the innovation was a new kind of action: large-scale, muscularly controlled motion.

With this transformation of animal life in our sights, let's think for a moment about the mind-body problem in the background. Ordinary ways of thinking furnish us with a number of concepts that help us get a handle on what minds do. One is the concept of *subjectivity*. This concept arrives in a complementary pair with another: *agency*. Subjectivity is a matter of seeming, of for-me-ness. It points toward experience as something that happens to a person. Agency is a matter of doing, trying, initiating. Agency is by-me-ness; it is being a source of action and its effects. It points toward the things a person makes happen. Interestingly, the word "subject" (though not "subjectivity") also has another set of connotations, in which a subject is a doer or initiator: subject as opposed to object. This is not the last time that these concepts will become entangled.

As everyday concepts, subjectivity and agency gesture toward different aspects of a person or animal, a more sensory side and a more active side. From an evolutionary point of view, however, these are closely tied together. Sensing has its raison d'être in the control of action. Nothing is gained biologically from taking in information that is not put to use. The evolution of the mind includes the coupled evolution of agency and subjectivity. But not everything has to develop in lockstep. There might be, at some stage, a breakthrough in the particular realm of action. A new kind of agency might come into being alongside simpler sensory capacities.

All through this part of the book, I am influenced by the thinking of the Dutch psychologist and philosopher Fred Keijzer, and his emphasis on the *shaping of action* as a central concern in the early evolution of nervous systems. The discussions in this chapter of the formation of action on a multicellular scale, the size and importance of that achievement, and its relationship to the animal body have all been influenced by him. In the area of the relations between sensing and action in very early animals, Keijzer has an interesting suggestion. He thinks that some new kinds of sensing may come "for free," or almost for free, as a consequence of the shaping of complex action. Suppose you build an elaborate system for producing some sort of coordinated, choreographed motion. Often, to do this you will need some parts of the system to be sensitive to what other parts of the same system are up to. But then if something external affects the system, especially by touch, this event will automatically be registered to some extent, as it will disrupt the pattern of activity going on among those parts. The internal sensing within the system will—or could easily come to—register that something external has happened. Even if a nervous system were to be looking entirely inward (Keijzer is not suggesting that things are ever like that, but if it were), that system would be responsive to things on the outside. Such a system can't help seeing outward, to some extent. New and expansive actions carry an expansion of sensitivity along with them.

The appearance of an asymmetry between complex action and simpler sensing at some early stage in animal evolution might be entirely an illusion. Complex sensing can be well hidden. But when thinking about the first forms of experience, or about what exists in animals *before* experience, it is interesting to imagine an animal whose motions outrun its senses, and ask whether, as Keijzer thinks, sensing would tend to automatically catch up.

Let's step back from these speculations to the main themes emerging at this stage of the story. All living beings do things. They adjust their activities, and affect what is around them. But in animals, this took a new form. On the animal outgrowth of the tree of life, one thing produced was a multicellular self. In addition, the evolution of animals produced multicellular action, action achieved through sheets of cells that contract, twist, and grasp. Nerves and muscle made this possible; a sponge can do nothing like it. Action of this kind was a transformative invention in evolution; it changed everything.

It changed everything—eventually. When did this transformation begin, and what sort of animal set things rolling? Was it an animal that looked like a cnidarian, or something else, earlier? As we'll now see, the engine of animal action, this engineer of Earth, had perhaps a fitful start.

Avalon to Nama

In the previous chapter we looked for clues about early forms of animal life, mostly by looking at animals alive now, but far from us. The outer limbs of the animal part of the tree of life, from our point of view, are very unclear. Things get clearer a few steps closer to us. Drawing evolutionary lines with time running up the page, the shape of some of the relationships looks like this:

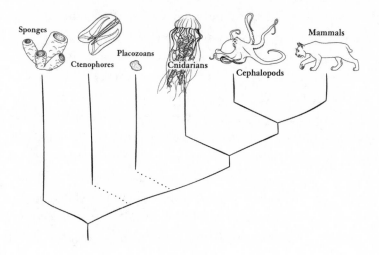

Nervous systems evolved somewhere below the split on the right that leads toward mammals and cephalopods on one side and cnidarians on the other. They may have evolved twice; this depends on unresolved questions about the tree, indicated by broken lines in the drawing.

All the branchings and evolutionary inventions discussed so far occurred before any fossil record of animals appears. The first period of time from which we have definite fossil evidence of animals is the Ediacaran, which began about 635 million years ago. The slow opening of the curtain on an animal fossil record reveals a scene very different from life around us now.

The setting is the seafloor, sometimes shallow, sometimes deeper, populated by various soft-bodied creatures, some tiny but some up to several feet across. Despite their soft bodies, a number of them left traces. Those traces show enigmatic forms—flower-like designs, whorls and discs, spirals and fractal branchings.

What reason is there to believe they were animals at all? In some cases this is indeed uncertain, and some remains may represent a wholly lost multicellular experiment, or experiments, far from the animals. But in at least some cases, they are animals. This was confirmed in 2018 by a student named Ilya Bobrovskiy, rappelling down a cliff in a remote part of Russia where unusually large and well-preserved remains of a famous Ediacaran organism, *Dickinsonia*, are found. As Bobrovskiy suspected, the rocks contained not just ordinary fossils, but remains that had been naturally mummified and preserved for over half a billion years. The mummified bodies contain cholesterol, a chemical made only by animals. *Dickinsonia* was flat, lived almost certainly on the seafloor, and looks like a bath mat, up to a meter long. It has no sign of eyes, limbs, or other familiar animal parts, and that is how it is with Ediacarans. They often have definite body forms—fronds and wheels, three-sided and five-sided—but no legs, fins, or claws, and no sign of complex senses like eyes.

No clear cases of cnidarians or sponges, the animals I've been discussing as clues, are known from the Ediacaran, either. But in each case there are possibilities. Some Ediacaran organisms look a fair bit like present-day "sea pens." These well-named organisms are in the same group as the soft corals we visited at the start of this chapter. But rather than trees, each looks like an old plumed pen with its nib stuck in the seafloor and feathers spreading on a stem above.

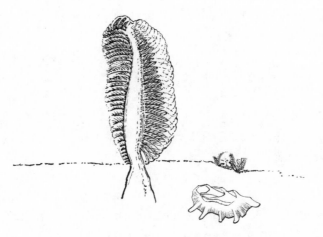

Whether any Ediacarans are closely related to sea pens remains controversial, as differences appear once one looks closely. Other Ediacaran organisms have branched fronds that also suggest cnidarian forms, but those similarities, too, may be misleading. A lot of Ediacarans were initially called "jellyfish," by Reg Sprigg, who first uncovered Ediacaran fossils while inspecting disused mines in South Australia in 1946. Most of those fossils have now been interpreted differently, but up in the water were real jellyfish, most likely, their bodies scrunching into incomprehensibility when they fell.

The Ediacaran in the biological imagination has often seemed a quiet and placid time, a period with little interaction between organisms. There are almost no signs of predation—no half-eaten individuals, no sign of the built-in weapons, offensive and defensive, that animals tend to have now. There are no claws or spines. No signs are known of sexual specialization, either, though it is hard to tell, and no Ediacaran has ever been assigned to one sex or another. Sex was almost certainly present, though it probably coexisted with various kinds of asexual reproduction (as in cnidarians and sponges today). Densities were often high; one can see slabs of rock with a jumble of dozens or hundreds of organisms of several species. But even in these Hieronymous Bosch–like scenes, the animals are not obviously having much to do with each other. There may have been hidden interaction by means of lost soft parts, but much of the familiar machinery of contact between animals seems to be missing.

That placid image of the time is fine as far as it goes. Recent years, though, have seen the emergence of more detail, and the quiet Ediacaran has begun to resolve into something with a little more drama, certainly with transitions and changes.

As it looks now, there were three stages. These were distinguished by a young biologist, Ben Waggoner, about twenty years ago, and have stood up well as more data has come in. The stages have lovely names (thanks to Waggoner, with help from the geography). I say "stages," but technically each is an "assemblage" (not as lovely a word); an assemblage is a collection of species as fossils, all laid down at roughly the same time.

The first of these assemblages is the Avalon, dating from about 575 million years ago. Even this first stage is fairly late in the Ediacaran as a whole. That period is bounded at its far end, about 635 million years ago, by the easing of an ice age, a massive glaciation that may have covered the whole Earth in ice. Conditions were quiet for a time, another ice age passed, and then the fossils appear soon after that. Oxygen levels may have increased significantly after this second ice age. Through all these early stages, however, one should imagine a low-oxygen world. This would have limited animal activity, to the extent that activity was even an option.

The Avalon assemblage, named after a place in Canada, features what seem to be stationary, plant-like, frond-ish organisms. (In a fortunate etymological confluence, "Avalon" in ancient Welsh means "the island of fruit trees.") A lot of these creatures look superficially like a single large leaf, or a cluster of leaves, stuck in the seafloor. Inspected closely, each leaf is a sheaf of intricate branching elements.

The Avalon assemblage also has a possible sponge—conical, with the right sort of shape, though not reminiscent of any modern sponge group. Sponges in general are puzzling. Chemical evidence along with genetics suggests that sponges were present, even common, at this time, but fossils so far give us just this one conical candidate and one later possibility, recently discovered, that looks like an old downturned TV antenna, with thin rods coming out from its center.

The Avalon organisms seem to have lived in quite deep seas, too dark for photosynthesis, hundreds and perhaps thousands of meters down. That is today a sparsely populated and challenging zone, but it may have been a cradle for slow but innovative life back then. These creatures might have lived on tiny dissolved particles of organic carbon—their branching-upon-branching design has a "fractal" organization that maximizes surface area, allowing a steady uptake of this organic mist, along with oxygen to burn it.

Next came what looks like a shift. The White Sea assemblage, named for a location in Russia, dates from around 560 million years ago. These fossils show more diverse body plans. There are still no fins or legs, but in a few cases, the body plan and fossil traces strongly suggest that the animal could move.

The creatures of the White Sea assemblage lived on shallow seafloors, not the depths as in Avalon. These are seafloors that were, in a sense, alive. They are sometimes referred to as "microbial mats," but Mary Droser of the University of California, Riverside, who has become a central figure in work on this stage, calls them "textured organic surfaces." These surfaces contained more than bacteria and the like, probably including algae-like organisms and small embedded animals. The fossil record preserves the textures themselves—"wavy-crinkly lamination, elephant-skin textures . . ." A tangle of living and dead organisms at different scales made up a mostly two-dimensional surface, a flatland in the sea.

In this setting, new bodies and lifestyles are seen. There are still stationary organisms, raised up and sea-pen-like, but also a range of flatter forms, oriented to grazing on the mat. Some of these could move. *Dickinsonia* (with its mummified Russian cholesterol) seemed to graze in one place, then move on, leaving a series of faint whole-body footprints behind. Two others were more active, it seems. *Kimberella* is seen as a likely relative of molluscs. It looked like a crawling macaron, one able to scrape the mat's surface with an extendable spade-like part.

There is also an enigma called *Helminthoidichnites*. This fossil, christened with that unbelievably difficult name (Hel-min-thoid-ich-night-ies) in the nineteenth century, had initially been found in less ancient rocks, and interpreted as the tracks of a small burrowing animal—perhaps a worm or crustacean. Similar traces were eventually discovered in Ediacaran rocks, and have now been analyzed in detail by Mary Droser and Jim Gehling, in South Australia, near where the very first Ediacaran fossils were found.

This work has used a new kind of excavation, enabling the undersides of huge rock slabs to be studied as wholes. Scrutinized closely, some slabs show traces of complex patterns of movement. The living animal moved through different layers of the undersea mat, piling up levees as it went. The tracks home in on the bodies of other animals, including *Dickinsonia*. This is the first fossil evidence of scavenging, of an animal consuming the dead. It is also the first physical trace of targeted motion, movement aimed at a goal that is sensed. Initially the targets were dead bodies, but scavenging has natural transitions to predation, especially when the prey is stationary or slow.

I said that *Helminthoidichnites* was an enigma. All Ediacaran organisms are enigmatic to some degree, but in this case the enigma is extreme. For a long time, all we had was the trail, and no remains of the animal itself. Just as this book was being finalized, a candidate appeared, a tiny bean-shaped creature that might be the maker of the *Helminthoidichnites* traces. It was found in South Australia, the original Ediacaran home.

The period of the White Sea fossils, then, is marked by a shift: new bodies, new behavioral capacities, a different environment. A couple of other animals from this period also seem to have been mobile. *Spriggina* has a body with movement written over its form, superficially resembling a scurrying trilobite. There are no known traces of *Spriggina* movement, but that is not a surprise, as an animal must burrow or scrape in order to leave a mark. If you just glide over the top of the mat, no trace will come down to us, many millions of years later.

This was a time when oxygen levels continued to rise, slowly and erratically. Perhaps the sequence is something like this: with more oxygen, textured living surfaces developed. They became a resource for grazing, encouraging slow movement along the mat. Grazing leads to a concentration of resources in animal bodies, which then die. This makes the environment more patchy—there is lots of food here, less there. Motion becomes more valuable, as does the ability to follow scents in the sea, to track things down.

The third named stage, after Avalon and White Sea, is called "Nama," after a site in Namibia, Africa. This is the youngest, leading up to the end of the Ediacaran. Given what had been going on just before, we might expect that the Nama phase would have more of this crawling complexity. But instead, these rocks are quieter. The crawling forms, surprisingly, are gone. The *Helminthoidichnites* traces are still there, and one interpretation of this period, called "Wormworld," holds that a profusion of small tunnelers and burrowers were major players in this third stage. But those larger mobile animals that look at least dimly reminscent of molluscs and the like seem to have disappeared. Other than the burrowers, life in Nama moved back to swaying frond-like forms (though with fronds mostly different from before). No one knows why this happened. And the Nama assemblage seems to represent the stage leading up to the Ediacaran's end.

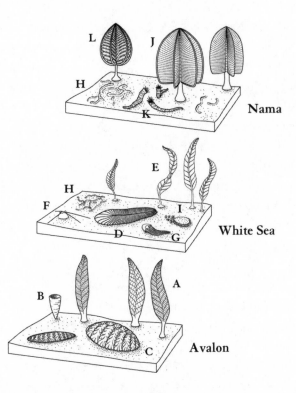

Three stages of the Ediacaran. Organisms include **A:** *Charnia*; **B:** *Thectardis* (a sponge?); **C:** *Fractofusus*; **D:** *Dickinsonia*; **E:** *Arborea*; **F:** *Coroncollina* (also a sponge?); **G:** *Spriggina*; **H:** *Helminthoidichnites*; **I:** *Kimberella*; **J:** *Swarpuntia*; **K:** *Cloudina*; and **L:** *Rangea*. *Charnia* and *Rangea* are some of the organisms that have been compared to sea pens.

How does this fit in with the themes in this chapter, with the attempt to discern clues about the evolution of animal action? We have a stage with stationary plant-like forms—Avalon, in the deeper sea—and then a transition to mobile creatures in shallower seas. Genetic evidence suggests that nervous systems evolved before any of these fossils were laid down, or perhaps during the first stage; the genetic dating is rough. Then there seem to be the beginnings of a regime in which new kinds of sensing and acting come online—White Sea. In Nama, this appears to fade.

If the Avalonians we know of were animals with nervous systems (not counting the possible sponge), what were they doing with them? It is tempting to say they might have been coordinating a reach or clasp, as in soft corals today. But in some cases where these organisms left unusually detailed fossils, no evidence of body openings has been found—unlike a soft coral, there is no mouth to bring food to. Instead, their whole body surfaces may have absorbed food—this, again, would make sense of a body design that maximizes surface area.

The frond-ish Avalonian organisms may not have been animals at all. Even if so, nervous systems probably did exist in some form before the crawling animals seen later in the Ediacaran. The evidence is good that nervous systems evolved in a body with something like a radial design. They might have been hidden in some of the flower-like bodies, but we might also need to look *up*.

I have been focusing on these seafloor forms because that is where fossil evidence was left. But all sorts of additional life may have been aloft in the water column—soft-bodied swimmers, akin to jellyfish and comb jellies. Some early stages in the evolution of nervous systems may have occurred up there. Filmy early swimmers will often be invisible to paleontology, as their delicate bodies are unlikely to fossilize into visible forms when they die. It has been suggested that early crawling animals might have evolved from a larva (an immature form) of a cnidarian-like animal, a larva that reaches the seafloor and starts to move along it.

From there one can envision a sequence from mat-grazing and movement to behavior directed on other animals.

Discussions of the Ediacaran sometimes seem to show an understandable preference for making the story work using just the cast of characters present in fossils, and hence living on the seafloor. I often wonder whether, with respect to the evolution of behavior and interaction between animals, the scene on the ocean floor is just the tip of an iceberg, and a lot of what matters took place up in the water column, with animals that left little or no trace. If that is true, it's hard to work out how the missing pieces of the puzzle might be filled in. But in evolutionary biology, it's remarkable how often the apparently unknowable becomes knowable, as a result of a sudden technical advance, a mummified *Dickinsonia*, or a new theoretical idea.

This talk of a transition from swimming to crawling also brings up a stage that has gone unremarked upon so far, but a development we should look at closely. Peering back into the undersea fog of those times, one of the events that had enormous downstream consequences was the evolution of a new kind of body, called the *bilaterian*, or bilaterally symmetrical, body. These are bodies with a left-right axis, as well as a top and bottom. Our bodies are bilaterian bodies, along with those of ants, snails, and seahorses. We have arms and legs on each side, also eyes and ears; there is a left-right pairing of many of the body's parts. Most animals now are bilaterians (that is true however you calculate "most"). So are some of the ancients—*Kimberella*, *Spriggina*, and others. Cnidarians like corals and jellyfish are not like this, and neither are comb jellies, sponges, or placozoans.

This body form evolved before the Ediacaran's White Sea stage. It had to be as early as this, because some form of this design had to be in place before the branchings that sent different bilaterian animals down their various paths, and there are at least a handful of different bilaterians in the White Sea. The last common ancestor of you and a butterfly, also the last common ancestor of you and an octopus, lived at least this long ago.

Traction

The bilaterian body, with its left-right symmetry, was an innovation especially in the realm of action. The bilaterian body is set up to *go* somewhere. There are no non-bilaterian animals at all on dry land—no crawling or walking jellyfish, no anemones with their fingers out in the air (though there are some that live in the intertidal). Bilaterian bodies seem to have begun on a marine version of the land: the seafloor. They are bodies made for crawling over surfaces, with direction and traction.

The very first bilaterians were probably simpler than the fossilizing White Sea Ediacarans, and possibly small, but we don't know much more than that. Once again there is a clue living today—another clue that might echo old forms. The clue in this case is flatworms. These are small, simple animals whose name does indeed capture how they look.

How useful a clue are flatworms? Perhaps not very useful. Flatworms today, no matter how simple, have been going about their business for a long while. The flatworm body form seems to have arisen, in some approximation, a couple of distinct times, and it makes sense that it would. Many flatworms today are parasites on other organisms, and parasitism often makes for simplicity. Flatworms may not, then, be especially good models for early bilaterians. But with all this in mind, let's take a closer look at them, as animals in their own right and as possible echoes of something older.

You can find marine flatworms on reefs and on fields of underwater debris. The ones that are noticeable are not the simplest ones—*acoel* (pronounced "a-seel") flatworms, which have been seen as the best clues for an early bilaterian form—but another kind, called *polyclad* flatworms.

They do look simple. They are about a centimeter long in many cases, though sometimes a good deal larger or smaller. They are often oval-shaped, flat with ruffled edges, and very thin. They look like a bit of tissue paper.

If you pause, though, after a few moments it is clear that they have a lot going on. They move faster than a lot of other animals down there. Some flatworms can swim, but even those that crawl often do so briskly and purposefully. "I might look like a bit of tissue paper, but right now I really have *something to do*."

Their bodies achieve a lot with a little. There is no "through gut"—what goes in one end goes out the same end—and no circulatory system. A cyclops-like eye cluster sometimes sits in the middle of the back, though eyes, of a very simple kind, seem to pop up just about anywhere on these animals, including on tiny pinches of the tissue-like body, raised a little above the plane of the rest.

Flatworm sex lives are more complicated than one might expect. They are generally hermaphrodites, and some engage in "penis fencing," trying to dart sperm into the body of another, while the partner does the same thing. It also comes as a surprise that they are often rather exquisite creatures, with bright colors and patterns. Initially, it is not at all clear why this is; they can't see each other with their simple eyes. But quite a few flatworms are mimics, especially of other small crawling animals, called *nudibranchs*.

Nudibranchs ("noodi-branks") are slugs, hence molluscs. They are closely related to slugs and snails on land, but can be astoundingly beautiful, with an endless range of colors and patterns. They often feed on chemically challenging food such as sponges, making them unpalatable to fish and other predators. The bright colors probably advertise this fact.

Two of the main groups are the *dorids* and the *aeolids*. The former are helpfully slug-shaped, from the point of view of a flatworm attempting mimicry, while the latter are named for appendages all over their bodies that look like streamers, continually moving underwater as if caught by a breeze. (In Greek mythology, Aeolus was the keeper of the winds.) Each spring, on a reef near the site of this chapter's opening, tiny aeolids appear. They are found especially on bryozoans, shrub-like animals made from tangles of strands. Aeolids appear on them, almost impossible to see. They are like tiny jeweled birds, millimeters long, in trees with branches like fine hair.

Near to aeolids on the evolutionary tree is *Tritonia*. This group of species has some large and well-studied animals, but also an elusive form that lives in this chapter's soft corals and sponges. They are very small, pearl white, with pointed outgrowths from their bodies like spires. On these extended outgrowths are further outgrowths, spires on spires. They look like they were designed by a miniature architect, a tiny Antoni Gaudí.

Those white Tritonias are hard to see, I said, and this is partly because they blend in with soft coral polyps, with their pearl-white spire-on-spire extensions. It took me a while to realize that this, too, is a plausible example of mimicry. I once saw a Tritonia moving about very slowly near a soft coral colony. For some reason I found myself thinking not of mimicry or camouflage, but of homage. The two had taken on the same shape, and stood near each other in a sort of harmony of form, with the coral reaching and the Tritonia appearing to follow: "My body," it seemed to say, "pays homage to the first animal actions."

THE ONE-ARMED SHRIMP

Probably a crab would be filled with a sense of personal outrage if it could hear us class it without ado or apology as a crustacean, and thus dispose of it. "I am no such thing," it would say; "I am MYSELF, MYSELF alone."

—WILLIAM JAMES, *The Varieties of Religious Experience*

Maestro

A little way from the scene that opened the previous chapter—the sandy plain, fast tides, and soft coral ascending—an old slender pipeline runs along the seafloor, from the shore out into the bay. A jumble of life has grown up around the half-buried pipe. As you swim along it, sponges imperceptibly adjust the flow of water through their bodies and soft corals unfurl their finger-like tentacles. You move along past slow forms of animal action. And then, under a ledge, something quite different: a movement of feelers. The fussing of many legs. And attention, orientation toward you, as you approach.

Banded shrimp have bodies with red and white stripes like old barber's poles. They are crustaceans, like lobsters and crabs. The term "shrimp" is a loose one that does not pick out a single branch of the evolutionary tree; the animals commonly called "shrimp" are collected from several nearby branches. Crustaceans, for that matter, are not a single branch, either. But all these animals are arthropods, and that *is* a branch of the tree, a large and significant one. Banded shrimp (scientific name *Stenopus hispidus*) have bodies a few inches long, two large arms with impressive claws, and several long white feelers or antennae, longer than their bodies. I saw a pair of these animals together at a particular spot, and months later, when I was nearby, I went to see if they, or others like them, were still there.

I turned up at the spot and saw a banded shrimp sitting on an ascidian spout. This shrimp had lost one of its big clawed arms. It was not greatly shortchanged, as it had a profusion of other limbs, including several smaller claws to accompany the remaining huge one.

The shrimp spent some time standing in front of a small dozing shark under a rock ledge, seemingly poking at it. The shark shifted a bit and paid little attention. Then, at one point, the shrimp climbed up under the ledge and must have hung upside down, as only a few long feelers like a cat's whiskers were visible coming downward.

I thought: why not touch them? It might scare the shrimp, but it could flee farther under the ledge if it wanted. It seemed entirely unconcerned about the shark. So I reached out and gently touched one feeler, with a stroke. To my considerable surprise, the shrimp climbed straight down and looked back at me.

I was delighted. After years following octopuses, I am now used to the idea of some degree of contact with implausible animals, but I was taken aback by the look full in the face given to me by this shrimp.

It resumed some wandering about. What was going on inside it? It registered me and allowed further touches. It stood with that one arm raised like a conductor, a tiny maestro, in front of the dozing shark.

Of the animals described so far in this book, this is the first one that can *see* an object like you, the first one with eyes that can distinguish objects (though some unusual jellyfish I mentioned earlier, the box jellyfish, do have eyes approximating this kind). This is also the first animal I've described that can move *fast*. It can climb rapidly, toward you or away. It can manipulate things. It can direct action on objects, as well as directing its gaze. This is an animal with a different relationship to its surroundings, a different mode of being, from any others discussed so far. How did this come about?

The Cambrian

Animals of this kind are a product of the Cambrian, another pivotal time in the history of animal life. In the previous chapter we looked at the period just before it, the Ediacaran, the time of the first known animal fossils. This period took animals from flower-like stillness to crawling and burrowing. The Cambrian, beginning about 540 million years ago, seems to see a sudden shift, almost a rupture. Among its fossils are animals with hard parts, legs and shells, and conspicuous eyes. The pioneers were old relatives of that shrimp.

Whether the appearance of a *sudden* shift is misleading is still debated (and "sudden" would still involve a million-year time-scale). There are several views about why a very different range of animals appeared at this time, though the views often invoke factors that can operate in tandem. Environmental conditions had changed, with more oxygen available. The ocean chemistry was becoming more encouraging of animal life. But also, perhaps enabled by the oxygen but going beyond any merely chemical effect, a new regime began in evolution itself.

Back in the Ediacaran, in the White Sea phase, we saw animals beginning to creep over surfaces and burrow a little way into the undersea swamp. We also saw the appearance of scavengers. An evolutionary pathway becomes visible, one that begins with moving slowly toward food, and then not so slowly, as others may get there first. From there, scavenging morphs into predation. This produces an "arms race." Once animals are trying to eat you, tracking your scent and following, it pays to improve one's own senses and means of motion. The evolution of eyes may have had a special role in this process, but soon it encompassed many facets of sensing and behavior.

The neatest fossil tableau of this process might include the remains of outclassed Ediacarans, mauled by the newcomers. In fact, there is no fossil record of an active replacement, as the Ediacarans have disappeared, as far as we can tell, by the time Cambrian animals arose. They seem to have just drifted offstage and been replaced by a different cast.

Arthropods appear to have led the way. This group today includes vast numbers of insects—those are later arrivals, and we will get to them in time. The earlier forms often look recognizably like crustaceans, though the most conspicuous ones, trilobites, may be more closely related to spiders. Arthropods at this time seem to have invented a new way of being an animal, with a skeleton that scaffolds and organizes complex actions. They also invented claws, and to go with them, image-forming eyes.

The banded shrimp exemplifies this new kind of animal; it is the very essence of arthropod, encapsulating their way of being. Amid the fidgeting of all the limbs, it took me a while to work out what was there. Basically, there are two big claws on long arms—except on the individual I was visiting, which had lost one. There are two pairs of smaller claws or pincers, with grasping devices at their ends. That is four additional claws. Then there are some legs, various things that look like extendable combs, and other minor parts. Looking at photos, the usual toolkit seems to be something like six claws, four legs, six feelers, and two other comb-like bits of paraphernalia. That makes eighteen limbs and protrusions, at least. A body like a Swiss Army knife.

This is not an easy body to keep track of. At one point in the events described above, the shrimp's one big claw found itself starting to grab one of its own legs, and quickly backed off, freeing the wandering leg. It used various of these appendages, especially the smaller claws, to rummage about and pick up bits of food. These actions are a long way from the slow unfurling of the soft coral and the silent pumping of the sponge.

This emptied-out-toolbox appearance is fairly typical of arthropods. If we are building a hermit crab, *why not* put some things that look like spatulas on the top of the face? Why not? That is the arthropod way of evolving: when in doubt, add some legs. Add some spatulas to your head. My one-armed shrimp was from a species that usually has two, but there are some crustaceans that naturally have one small claw and one huge, outsize one, as if the animal bought it late at night online.

Arthropod evolution has been exuberant for half a billion years. The largest arthropod that has ever been found is a giant *anomalocarid*. This group, first appearing in the Cambrian, featured swimming predatory forms. The largest, though, at seven feet long or more, was a peaceful plankton feeder, somewhat like a baleen whale.

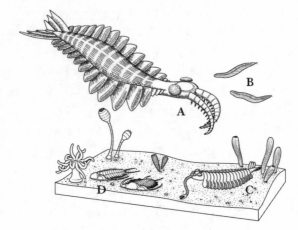

A Cambrian scene. The animals include **A:** *Anomalocaris* (a predatory one, not the gentle giant mentioned above in the text, who came later); **B:** *Pikaea*, whom we will meet in Chapter 7; **C:** *Opabinia*, another predator related to arthropods; and **D:** *Cheirurus*, a trilobite.

Not as many arthropods swim now—their competition in open water has been different, and formidable, for a long time. Today's swimmers are often small forms, and often delicate and beautiful. Diving in Lembeh Strait, Indonesia, I saw swimming anemone shrimp of the genus *Periclimenes* (named, I think, for a shape-changing grandson of Poseidon, the Greek god of the sea, though many species are now being moved into the genus *Ancyclomenes*, named from a Greek word for "bent" and sounding like an antibiotic). These shrimp are cleaners of other animals, but live among anemones for protection. They are tiny, almost transparent, with a few splashes and stripes of bright color.

Buzzing around their anemone, they looked like a happy group of angels.

In these animals, with their good eyes and finely engineered claws, there is engagement with external objects, on both the sensory and active sides. They can see objects and manipulate them. Though the banded shrimp I visited had lost an arm, it was the only animal in that scene with anything like arms or claws. Around it there were molluscs and worms, also the shark, and none of these have much in the way of means for manipulation. (There were no octopuses around that day.) The arthropod could act on objects in ways that no one else down there could match—truly a maestro. That is how things were back in the Cambrian, and unlike the sleeping shark, no animals back then had teeth. In the land of the limbless, the one-armed shrimp is king.

Animal Sensing

Chapter 3 looked at the early history of animal action. Now, under the attentive eyes of the shrimp, we turn to some episodes in the history of sensing.

Sensing, like action, was not invented by animals; sensing of some kind is seen in all known cellular life. Single-celled organisms can track touches, chemicals, light, and even Earth's magnetic field. But in animals, sensing saw a transition—it saw several, in fact.

In the previous chapter we looked at action that involves the coordination of many parts, organized in space. The same sort of change occurred with sensing. Animals evolved surfaces, arrays, and screens that are sensitive, that can take an imprint or image.

The retina in one of your eyes, for example, is an organized layer of cells on which incoming light forms a pattern. The pattern or image is not *seen* by the brain, but its spatial layout—which bits are next to which—has effects downstream, on the neurons connected to it. The sensors in our skin that detect touch are similar, registering the shape of an imprint or the texture of an object.

Earlier I raised the idea that the evolution of action "ran ahead" a little of the evolution of sensing. That was a speculation. We don't know what sort of ordering there was between the evolution of new kinds of sensing and new kinds of action. And once we are in the Cambrian, both advanced rapidly. In any case, just as animals became new kinds of objects when they evolved multicellular action, they did this also when they evolved multicellular sensing, encouraging parts of their bodies to become maps or reflections of fragments of their surroundings.

Eyes are a paradigm case, and they became sophisticated early in the Cambrian. Arthropod eyes are in most cases "compound," with lots of small parts that each have their own lens. Our eyes, in contrast, each have one lens and retina—this is the "camera" design. Some spiders are special cases among arthropods, as they have eyes more like ours (and some have a telephoto lens inserted as well). But the most elaborate arthropod eyes, and by some standards the most elaborate eyes of any animal at all, are those of the mantis shrimp.

Mantis shrimp, or stomatopods, are among the most active marine arthropods alive today. They are not very big, but in some ways they recall Cambrian years of arthropod dominance. I followed one of these for a while, also in Lembeh Strait in Indonesia. It was about six inches long, looked like a compact lobster, and in true arthropod style had a head festooned with golf clubs and party lights.

It scooted away across the seafloor. I came after it, not too fast, but persistently. It kept moving and every now and then would stop suddenly and look back. I imagined it saying "WHAT? WHAT?" each time—an irritated interrogation. (A more accurate interpretation might be "check . . . check . . .") One would normally expect an animal doing this to turn its head or body back to look, but a mantis shrimp's eyes are two spheres on stalks that can rotate freely and do so independently of each other. It could do its irritated inspection without turning around.

Mantis shrimp can look at the same object with different parts of one eye, and thereby see depth with that single eye. Their eyes have about a dozen different color receptors, as compared to our usual three. These animals also have elaborate weaponry. The armaments they sling are basically hammers and pikes, organized with "springs, latches, and lever arms" (I am quoting from a paper coauthored by Roy Caldwell, the Berkeley biologist and stomatopod whisperer who has done more than anyone to discern what these animals can do). When a hammer is released from this apparatus, its motion reaches, for a moment, extraordinary speeds and can vaporize water.

A banded shrimp has a more peaceful demeanor. In my encounters with them, when they came out to look, I had the initial impression that their response was inquisitive. Later I came to appreciate the biology of it. These shrimp, like the tiny ones I saw in Indonesia, are cleaners of larger animals. They eat parasites from the bodies of others—of fish, eels, and turtles. The shrimp greeting me in the opening pages of this chapter was probably coming down to check out a cleaning candidate, a client.

We now have in place both action and sensing on a multicellular scale. This is not just a sophistication of each side, a pair of complementary improvements. The combination gives a new shape to the relations between those two older capacities, sensing and action.

An ideal example with which to introduce this theme is the banded shrimp's feelers, or antennae. These are long, several times the length of the body itself. They are active, waving about in all directions, and also sensitive. If I reach out my finger and a feeler touches it, often the animal responds immediately. But these animals are frequently roaming around under a confined ledge. Their feelers must bump into things all the time, as a consequence of the animal's own actions. The same touch can have different causes, with its own actions being one such cause. The shrimp seems able to register which touches are just due to its wandering, and which are due to the actions of an *other*, such as me.

I don't know of research on the use of these feelers themselves, and on how banded shrimp manage the relationship between self-caused and other-caused events in this case. Animals of this general kind, though—crayfish and flies—have been shown to have systems which modulate the interpretation of sensory information by the animal's registration of what it is presently doing. This ability is a general feature of animals—not of all animals but a great many, including some with simpler nervous systems than arthropods'. It is a form of coordination between the parts of an animal that sense and the parts that act. You work out what sensory changes you would expect to encounter just as a consequence of what you are currently doing (whether you are moving or still, and so on), and look for sensory changes over and above these. The "over and above" changes are marks of something going on outside you, like someone reaching out and poking your antenna.

If an animal does not do something like this, its own motions will confound its attempt to understand what is going on. If an animal does do this, it is now sensing the world in a way that tracks the divide between *self* and *other*, between the animal itself and everything else. This tracking can sometimes be achieved by a nervous system in a very simple way, but however it is done, the animal is now doing things aimed at referring some events to external goings-on and other events to the self-caused. It is dealing with the world in a new way, sensitive to the distinction between external world and self.

A lot of literature describes this as the handling of a problem that animals face. I did that above. The neuroscientist Björn Merker sets things up this way in an influential paper. Moving around is good, Merker says, but moving has costs or "liabilities," and one of the costs is that the world becomes more confusing. But there is another way to look at this situation. The fact that your actions affect your senses is not just a problem; it is also an opportunity. By means of action you can probe the world and bring about new stimuli. You can poke and interfere and make the world talk back to you. The shrimp's feelers make that clear. Action brings to the senses a combination of difficulties and insights, simple or complex according to how much you can do and how much you can sense.

This kind of sensing that marks the relation between self and other is an important feature of the animal way of life. It gives rise to a new way of being in the world. It involves the establishment of a point of view, a perspective, in a new sense.

So far I have been talking generally about "sensing," but this phenomenon differs a lot between one sense and another. In vision and touch, your actions have immediate and strong effects on what you sense. A slight movement of the head changes the entire visual field, and this would be very confusing if you did not register that you had moved your head. The same applies to touch. Hearing is very different, though: the effects of your actions on what you hear are certainly real, but often slighter. If you move your head as you listen, there are consequences, but the auditory world does not dramatically shift. Small motions have less marked effects on what is heard. Smell and taste, the chemical senses, are different again—they sit perhaps between the two poles.

With all the fossil uncertainties, it is not clear when these new ways of sensing the world started. None of this seems likely to have arisen in a jump. The Cambrian probably had a special role, as both eyes and new ways of moving seem to date from this time. Some Cambrian trilobites also had antennae. The banded shrimp, again, is an exemplar of these themes. Those shrimp are very three-dimensional objects, with their feelers and manipulating pincers on long arms. They take up a lot of space; their bodies have a lot of *presence* in space. The shrimp is the source and center of much probing and manipulation of surrounding objects. I think the experience a shrimp would have is of a very spatialized world, including a distinction between exactly what is *it* and what is *not* it—remember, from before, the momentary grabbing and release of its own leg.

Cleaner animals, by the way, tend to distinguish themselves in this realm. Cleaners have a fine-grained engagement with their surroundings, as they are dealing with the most complex of all environmental factors, other agents. The "mirror test" is a test used to work out whether an animal can recognize itself in a mirror *as* itself, for example by grooming or cleaning a spot on its body that can only be seen in the mirror. Very few animals pass this test. The only animal reported to pass a version of it, so far, that is not a mammal or a bird (and only very few of those pass it) is a cleaner fish.

An Inquisitive Crab

Crustaceans are often active animals with good senses, and also rather long-lived. They have been seen, and treated, as little robots, an impression their hard shell encourages. But there is more going on inside these animals than has usually been supposed.

Some especially significant work has been done by Robert Elwood of Queen's University Belfast and his collaborators. Hermit crabs are the crabs that collect and live inside shells left by marine snails. They walk about with the shell around them like a coat of armor or a movable house. Elwood and his collaborators have collected considerable evidence that these animals can feel something like pain. What is important here is not that they flinch or react to what seem unpleasant events, but that things they do suggest that what is going on is not just a bodily reflex, and something like pain is being felt.

Pain will come up often in this book from now on, so I will introduce some terminology. The word *nociception* refers to the detection of damage, along with a response to it. (It is pronounced "no-see-ception.") Nociception is very common in animals, but it often lends itself to interpretation as something like a reflex. So biologists tend to regard nociception as not sufficient for pain, and they look for markers of something more, something that seems tied to a *feeling* of pain. All these markers, in animals that can't report to us how they feel, are controversial to some extent. They include tending and protecting wounds, seeking out pain-killing chemicals (the same drugs that work in us, in many cases), and some kinds of learning from the good and bad consequences of actions. Elwood and his group have shown a kind of wound tending in shrimp, for example. When an antenna was touched with vinegar or bleach, the animals would groom that antenna and rub it against the wall of the tank.

Another kind of experiment involves looking for trade-offs. The idea here is that if something feels bad to an animal, and the animal is also reasonably smart, it might balance the badness of that feeling with various other benefits and costs that apply in that situation. This would be very different from a reflex response. Hermit crabs do make trade-offs of this kind. Elwood's tests use small electric shocks. These shocks can induce a crab to abandon its shell. That alone would not mean much. But it was also found that if a crab had a better shell, it was more reluctant to give it up—it would tolerate more shock before leaving. If there was a smell of a predator around, the animal was also less inclined to abandon the shell. Again, it would tolerate a shock that would otherwise induce it to leave its shell. All this suggests that for a crab, there is a range of events and possibilities that are seen as good or bad, and the pain of a small shock, though bad, is factored into a decision along with other considerations.

The resulting decision takes all of them (or several) into account. Some other findings in this work are also striking in the way they suggest the presence of feeling. Crabs who had been shocked and exited a shell would sometimes afterward carefully inspect the shell, apparently trying to locate the source of the problem.

This work on crustaceans was, as far as I know, the first on any invertebrate animal that produced reasonably good evidence for pain. The work is not decisive, as Elwood recognizes. One might question the tests used. Elwood's reply has often been that he is using tests that are taken to be good evidence for pain when they are passed by more familiar vertebrate animals. A person might then say, "If a shrimp passes the test too, it shows the test is no good." That reply is possible and there is no knock-down argument against it. But it obviously has a rather ad hoc character—it looks like a mere attempt to avoid changing one's mind—unless further reasons are given. This work provides a real case for viewing these animals as able to feel something akin to pain.

A common argument against crustaceans feeling pain, an argument used also about various other animals, is a rather bad one. The argument is that crustaceans don't have the brain areas involved in human pain. But as Elwood replies, crustaceans also don't have brains with visual areas that are anything like ours, though they can plainly see. Evolution sometimes builds a range of different structures that carry out the same function. That applies clearly to vision, and is probable also for pain.

Consideration of the welfare of crustaceans in most countries is basically at zero; there is literally nothing that might be done to them that causes concern, and boiling them alive is routine. Hermit crabs might be somewhat different from other

crustaceans—they do seem to lead quite complicated lives—
but the evidence for pain in crustaceans is not confined to them.
Crustaceans have capacities that people had not suspected.

I was underwater a while back photographing something
very still—an ascidian under a ledge that had a sponge encasing
it in an almost perfect sphere. It looked like a suspended purple
moon in dark space. I was there for a while quietly fussing with
the camera, and suddenly there was a great clatter of movement.
A large hermit crab came tumbling off the rock shelf and landed
right in front of me. The combination of house and crab was about
the size of an orange. As nothing at all had been going on just
before, I strongly suspect the crab had been spying on me from
the ledge and had overbalanced, resulting in a long, somersaulting,
head-over-heels-over-claws descent, crashing gently down before
me. She jumped up immediately and ran under a ledge.*

With some hesitation—I don't usually do this sort of
thing—I gently picked her up and placed her in the open. She
came rocketing back toward the ledge.

When I picked her up, a mass of bright-orange streamers,
like tiny fireworks, came pouring out. These are defenses, stinging strands called *acontia*, emitted not by the crab but by anemones. As well as adopting shells, some hermit crabs pick anemones

* This is the first of a number of times in this book where the best way to describe what
is happening with an animal is to refer to it as "he" or "she," in situations where it is hard
to tell the sexes apart and I am not sure what the sex of the individual is. (There were also
some hermaphrodites at the end of Chapter 3, but pronouns weren't needed in those
cases.) I am reluctant to call them all "it," and though in many contexts I favor the singu-
lar "they," it doesn't always work here. The stories do feel different when one assigns the
individuals one sex or the other. In these cases I will assign the animal a particular sex if
I have any clues at all to guide me, even if they are unreliable clues, and describe those
clues in the text or the endnotes. If I have no evidence, I will assign a sex arbitrarily based
on some quirk. In this first case, I have no real clues but an early study of hermit-anemone
associations noted that almost only females, among the individuals observed, picked up
anemones and placed them on the shell, rather than relying on the anemone to crawl onto
the shell on its own. The paper is cited in the endnotes.

up with their claws and carefully arrange them on the outside of those shells. They use the anemone's stingers to protect themselves from predators, especially octopuses. In some hermits, the odor of an octopus induces them to pick up anemones if they are under-dressed, and dominant individuals sometimes remove anemones from the shells of other crabs and put them on their own shell.

This crab, in any case, determinedly scampered back under the ledge, as far back as the shell would allow. Her eyes, on long stalks, glared out at me.

The Elwood work is important not only for crustaceans in their own right. A while ago, I had just got out of a scuba dive onto land, a dive surrounded by these engaged crustaceans, when it suddenly struck me that once one gets used to seeing crabs and shrimp as experiencing animals, this has ramifications for others, most markedly for insects.

In crabs and shrimp we have animals recognizable, even con-spicuous, in their marks of experience. They act at our pace and scale, and have our sort of agenda. Crustaceans are in the same large group of arthropods as insects—insects are probably an evolutionary offshoot that came out of a large "pancrustacean" group. On land we are surrounded by insects, and we kill them routinely and in colossal numbers without a thought. I have tended to think of insects as unfeeling robots, as most do. But there is a gestalt shift you can undergo by way of their relatives, the amiable crustaceans. The crustaceans put insects into a new light. Do insects experience their lives also?

This conclusion is not automatic. Life on land has taken in-sects down a different road. But it came as a bit of a jolt, standing on the shore, when I thought about it. We have to take insects seriously as candidates for subjective experience once we have the crustaceans and their story in place. Insects are physically smaller than crustaceans, in most cases, and don't act in such recognizable ways. But insects do not have simpler brains than crustaceans—many have considerably more complicated ones. Crabs show us in visible and graspable form what these kinds of animals can do, and what might be going on inside them.

Another Path

We are working toward an understanding of an *animal way of being*, one that is distinctive compared to other kinds of life. This way of being was brought about by the formation of the animal body and action, together with new kinds of sensing that accompany, and are fed by, those actions. Though there's a pattern to be discerned here, it is important not to oversimplify things. Alongside the stages in animal evolution we have been looking at in the last two chapters, another path was being explored, by other kinds of animals as well as by plants.

Imagine we have a small animal who might benefit, in an evolutionary sense, from getting bigger. There are two ways to do this. One is to retain the same shape and try to build that body on a larger scale. This brings about new demands for circulation of materials and coordination. The other approach is to just repeat the form you have, and then repeat it again. Add a twin to your body, a twin that remains attached. Then add another. This, in biology, is the *modular* body plan.

This process creates a tight-knit colony, a patchwork, of repeated units. The result is a bit like the situation with cells that repeatedly divide to form a body like ours, but now the repeated units are whole animals, or other similar units. This is how things work in corals, and also to a large extent in plants.

When you are looking at modular organisms, it tends to be unclear what counts as a single individual—is it the visible branching coral, or the individual polyps that make it up? The smaller units often have a fair degree of autonomy. They can often reproduce individually, for example, even if they depend on the larger unit to stay alive.

Modular organisms often produce branched, tree-like forms. Once you go down this path, your lifestyle in the realm of behavior tends to stay simple or become simpler. Corals, immobile though they reach, are an example. Others are in a sense more extreme.

Bryozoans, the shrub-like creatures we saw hosting nudibranchs in the previous chapter, represent an emphatic turning toward a plant-like form after a long period of time spent on the same evolutionary road as ants, octopuses, and other animals of that kind. Bryozoans (the name means "moss animals") are fairly closely related to molluscs. They have a bilateral left-right form and a nervous system. But they turned toward a different way of life in a very concerted manner. Many of them form colonies that look just like underwater bushes and mosses.

In organisms like this, especially those with branching, the resulting body has a form that is only partly predictable. An oak tree has a recognizable shape, but no fixed number of branches in the way a human has, with rare exceptions, a generally fixed number of limbs.

A human, shrimp, or octopus is a *unitary* organism. We have definite forms, repeated across generations, and are not built out of units that are partially self-sufficient. The unitary body form is important for the evolution of action: when there is a standard body shape, routines and patterns of action can gradually evolve. Nervous systems can fine-tune, generation after generation, the same coordinated motions.

Modular organisms, by contrast, tend to be stationary. A few can swim a little—a drifting swim. But when animals in the sea that are usually modular head toward motion, they tend to opt for a more unitary form. Swimming anemones, which can sometimes swim quite well, are one big polyp. Modular animals are not able to generate complicated actions as wholes, and in that way they are like plants.

Plants are far from inert, as we'll see in a later chapter. They can sense and respond. But for the most part they put these abilities to work in different ways from animals. Plants put them to work in the formation of the body itself. The shape of a plant reflects the history of what the plant has sensed—where it has found sun, and so on. As the body is less integrated, the form can be more variable, and freely suited to circumstances.

A while ago I was looking at a bryozoan colony underwater. Tom Davis, the diver from the soft coral time-lapse study, was showing me around a dive site, and he pointed out the colony. On the colony's stems were some tiny nudibranchs, each no more than a couple of millimeters long. The bryozoan was a tangle of translucent threads like glass noodles—the species is called the "spaghetti bryozoan." It looked nothing like an animal, or even a collection of them, but like a stationary spaghetti tangle.

I took a lot of photos to make sure I got the animals Tom was pointing at. Later, on the computer, I found the tiny slugs and also looked through the branches of the bryozoan, reflecting on how bush-like it was, with stems and nodes. These stems, again, were tiny animals with their own nervous systems, forever bonded to others. I wondered what was going on inside them. Then I saw an unexpected tiny stripe of red. Zooming in further, I realized the stripe was on a talon-like claw. The claw was at the end of what initially looked like just another stem of the bush, but that "stem" was attached to another one with something quite out of place: an unmistakable hinge. With claw and hinge I knew I was looking not at more of the bryozoan colony, but at an arthropod.

Its legs and segmented body were very thin and almost indistinguishable, at first, from the plant-like stems of the bryozoan. Soon I could make out its head and how the body was arranged. It was gaunt, almost transparent, with needle-sharp claws. I found another one of these, and a third. As a flip through different photos made clear, these creatures were in constant motion. They are called "skeleton shrimp." A collection of ferocious glassy skeletons was wandering through the bryozoan's branches. Both the pale stationary stems and taloned climbers were animals, with muscles and nerves, following different evolutionary roads.

I went back out to look for the skeletons a few weeks later. "Looking" was not easy, as you can hardly see them at all. I often found them only later on the computer, but when I did, I saw that they were *everywhere*. Crouching, hanging upside down, interacting with their tiny talons, clambering over the scenes I was photographing, they were there at the rear of the stage, barely visible. When I look at photos taken of other things, I now often see skeleton shrimp in a little horde in the background. They are like the crowded tiny ghosts of long-dead generations, haunting that bit of sea.

The Decorator

With all this, we have come a considerable way along the road that produced animal life. That road begins with some single-celled eukaryotes not much different from others, setting out on a route that led, later, to the integration of countless cells and a new kind of living unit. Through the origin of nervous systems and muscle, action is created on a multicellular scale. Then the first bilaterian animals arise, and from there a series of branchings. All this occurs before the Cambrian, and most of it before the first animal fossils. With the Cambrian comes an arms race of action and sensing. There we see a new mode of living pioneered by arthropods, with actions scaffolded by sculpted, standardized bodies and jointed limbs.

Here is another picture of part of the tree of life, picking up more of the stages and animals that have been discussed so far.

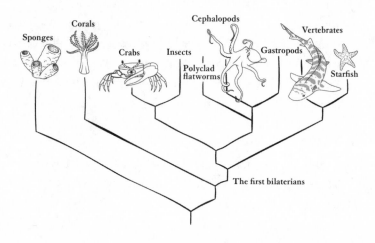

Again, time runs up the page, and a huge number of animals are not included. When a picture is drawn like this, the significance of the first bilaterians is evident. We don't know what these animals looked like, though flatworms have been seen as roughly indicative. As the drawing shows, a deep fork stemmed from some animal of that kind. Up to this time, your history was shared with the history of ant, crab, and octopus. After that, the lines evolved independently.

Diving one day on the undersea pipeline that opened this chapter, I found myself looking at a sponge—a mass of dark red with flat soft extensions. They looked like a collection of Dr. Seuss fingers. It was completely still, as a sponge is, to the unaided human eye. And then it suddenly creaked into motion. At the same moment, I realized there was a decorator crab there.

A decorator crab is another arthropod. It dresses up in sponge—encourages sponge to grow on its body surface, often producing specialized hooks on its body to attach the extra material. It does this far more completely than the hermit crabs who use anemones. A hermit climbs inside a snail shell and adds some anemones to the outside. A decorator crab uses its own body, its own shell, as a setting for the arrangement of other life. Often its bodily arboretum contains not just sponge but corals and other cnidarians. These are all stationary forms—and inedible ones, from most points of view. The crab is protecting itself. The decoration provides a combination of concealment and discouragement to many hunters, especially to an octopus.

Much of what had looked like the sponge was now moving. A claw appeared. These sponge-covered crabs move much more slowly than other crabs. Their motions are ponderous and slow, held back by the immobile sponge tissue over their bodies and joints. The head was slowly raised. It stood there embodying two very different products of animal evolution.

Goodbye

About two weeks after the time I visited the one-armed shrimp conducting the dozing shark, I went back to see if he was still there. I swam out and on the way I came across a rather aggressive octopus, who chased me away from her den. As I approached the shrimp's ledge, I worried that an octopus, perhaps that very one, might have eaten it during the weeks I'd been away. But there he was, a one-armed shrimp, hanging under the ledge. There was no sign of the shark.

The shrimp was less active, less interested in me than before, but eventually came down to stare and wave his arms. He looked a bit disheveled, as some algae had been gathering on his legs and body. I was pleased to see him, though.

During this period I read more about these shrimp. I learned that they are long-lived, territorial, and monogamous—mating for life. In aquarium settings they are listed as living for as much as five years. Banded shrimp can also recognize each other as individuals. In an old study from the 1970s, mated pairs were separated for a night or two and then reunited, and compared with shrimp who were separated from their mate and returned to another individual, a stranger. The strangers were of similar size and appearance to the lost mate (as far as humans could tell), as well as being of the same sex. The shrimp could tell the difference. Strangers, when brought together, both courted more and fought more. Mates tended to resume normal behavior.

This study reported that in the wild, mated pairs tend to stay within antenna's reach all day, but apparently wander (especially males) for a few meters at night and return before daylight. They seem to spend much of their lives in just a square meter or so of territory.

I thought about seeing a pair of shrimp at that spot months earlier. Both of them were very active, with no missing arms. I have video from that time, and saw the pair fussing intensely with their feelers, face-to-face. There was a lot of seemingly disorganized poking and touching. Was this signaling? Grooming? The 1977 paper decides that the detection of chemicals is the most likely basis for individual recognition in this species. But I wonder about the fussy many-appendaged face-to-face interactions I had seen in the pair.

Females are larger than males, but it is otherwise hard to tell the sexes apart. I could not determine the sex of my solo one-armed shrimp. I felt a bit sad about this animal, who was definitely missing a companion, whether or not he had been part of the first pair I'd seen there. As well as lacking an arm, he had a body showing the undersea signs of wear, strands of other life accumulating on it. This makes me wonder whether those face-to-face interactions include grooming.

Two weeks later I went back again, implausible as it seemed to be driving three hours up the coast to visit a shrimp. The water was not good this time. The whole place felt closer than usual, and darker. Ascidians coughed and sneezed. I arrived at the shrimp's ledge and there was a shrimp of the right species and size in the exact spot, alone. This one was missing both arms. It was probably the same one, having lost the other long arm.

Without even one long arm I initially wondered how he could eat. But he had started with so many limbs, and at least several of the four smaller claws remained. He seemed to be picking things up and shoveling food into his mouth with those sub-claws pretty well. He did seem to be doing all this in a less energetic manner than before, though. He looked tired, very much on his own, and probably near the end of his days.

5

THE ORIGIN OF SUBJECTS

Subjects, Agents, Selves

How far have we come? In evolution, the story so far has taken us from early life through early animals, then to the advent of action, nervous systems, and vision. We are presently parked at the Cambrian. There we are watching arthropods as especially active creatures, but in the background are the beginnings of animals that will take center stage in later chapters: vertebrates and also cephalopods, the inimitable molluscs that will bend our thinking about all these questions.

How far have we come on the philosophical questions, the project of making sense of the evolution of the mind and the relation between mental and physical? We have come some way; some of the ideas introduced in the preceding pages do begin to "bridge the gap." Others will come later, but this is a good place to take stock—a good place to make connections between the story that has been told so far and how things look from the philosophical side. This will involve a mixture of reaching out and pushing back: reaching out to show that there has been some progress, and pushing back against misconceptions that make the problem seem more intractable than it is.

What evolution has brought forth, in the story told over the last few chapters, is not just some new organisms, larger and more complicated than earlier ones, but a new kind of being, a new kind of self. These are organisms that are tied together in new ways, and have new kinds of engagement with their environments.

Central to animal evolution was the invention of new kinds of action, the coordinated motion of millions of cells, made possible by muscle and nervous systems. New kinds of sensing arose to guide these actions. With these innovations working together, we encounter beings that respond to the environment as a realm beyond, sensing and acting on the basis of a tacit sense of self (remember the shrimp who backed off from seizing his own leg with a wayward claw).

Two concepts I put on the table earlier are *subjectivity* and *agency*. These concepts pick out different aspects of a whole that is familiar from everyday life. That whole is a person, sensing what goes on around them and acting. Subjectivity involves feelings and seemings; agency is doing and initiating. All living things (or all living things composed of cells) exhibit something like subjectivity and agency, but these features take a different form in the animal case.

In philosophical debates about mind and body, subjectivity is seen as the main problem—"subjective experience" is one label for what is hard to explain. Agency seems easier to understand. But the sensing and feeling side of life, as we saw, is entangled with action in many ways. The philosopher Susan Hurley introduced a good image for thinking about these relationships.

Hurley said there is a standard picture that is not quite right. In that picture, a person is "seen as a subject and an agent standing, so to speak, back to back." A person becomes a partitioned or layered object, flanked by the world on each side. The world affects the person through the senses—that is the subject side. The person also works out what to do in response, and acts—that is the agent side. These become two subparts of a person, almost two people. But the roles are less separate than that, Hurley said, and the back-to-back image is misleading. We are subjects-and-agents together.

Here is another way to describe what arose, what was new. As evolution proceeded, animals became a new kind of intersection point, or nexus, in the world's networks of causal pathways. When an animal picks up information of various kinds through the senses, it becomes a point at which lines converge. When it is an initiator of action, it becomes a point from which causal lines *di*verge, spreading downstream, and often also looping back to affect the animal's senses. An animal is also a place where the present intersects with the past; your handling of the present will be affected by what you have seen before, and how well yesterday's actions seem to have turned out. Incoming information about the here and now interacts with traces of times past.

These features and this placement in the world are common, though not invariable, products of the animal way of life. Animals of this kind have a point of view, and from that viewpoint they act. Some of what falls under the problematic umbrella of "subjective experience" is an expected, comprehensible consequence of animal evolution. Roughly speaking, the evolution of animal agency brings with it the origin of subjects.

Qualia and Other Puzzles

In the first chapter I said that one of my aims is to remove a sense of arbitrariness in discussions of animal experience. One person will look at a paramecium and say it has feelings; another will not just dismiss the paramecium but also fish. I want to help us get past this situation. Some of what is missing here is information about particular creatures, about what they can do and what goes on inside them. Those are detailed scientific questions. But some of the difficulty comes from a kind of unmooring that can occur when we try to think about these things. This unmooring is related to a very old problem, one seen in traditional arguments against materialism.

René Descartes, in the seventeenth century, imagined that he might be a soul without a body, and argued as follows. Can you be sure that you *have* a physical body? It might be an illusion. On the other hand, you can't doubt that you have a mind (at least right now), so mind and body can't then be the same thing. You can't *be* your body and no more, if that body is something optional to your existence.

A more recent route is a reversal of Descartes's thought experiment. Consider the possibility of a body with no soul or mind. Specifically, imagine an exact physical duplicate of an ordinary person. If materialism is true, the duplicate ought surely to have experience. But this seems, again, to be optional. A physical duplicate of you might be a mere "zombie," as David Chalmers has put it, something wholly unconscious. If this really is possible, then it seems that body and mind can't be the same; the mind is something in addition. The mind might be something reliably made by brains, bodies, and physical processes, but it can't just *be* them.

Body and mind appear to be separable in this way, I agree. But this appearance of separability comes from a quirk of our imaginative capacities. Thomas Nagel, despite being a critic of materialism, diagnosed this quirk and indicated why it is misleading. Human imagination has several varieties, including "perceptual" imagining, which is imagining seeing or hearing something, and "sympathetic" imagining, which is imagining *being* something. Sympathetic imagination can only be applied to minds; you can only imagine being something else that (you think) has a mind, or at least has experiences. Perceptual imagining, in contrast, is most readily applied to bodies—to objects that can be seen, heard, or touched. In thought experiments, things that we sympathetically and perceptually imagine can be freely combined, separated, and rearranged. By means of one combination, we can imagine a soul with no body (we make the perceptual side "blank"); by another we can imagine a body with no soul (now the sympathetic side is blank). The fact that we can do this doesn't show anything about what can really be separated, what is genuinely distinct.

Nagel himself thought the mental is probably inexplicable in materialist terms, but he denied that we could show materialism to be false with thought experiments of this kind, and he was right to say that.

A similar unmooring is seen elsewhere, in milder forms. You can look at a paramecium and bestow it with an inner life, through an imaginative act of the sort Nagel described. You can also look at a fish and imagine it all dark inside. These imaginative exercises are uninformative when they are freely imposed in that fashion. Imagination does have a role, and in various places in this book I try to imagine our way into the lives of different animals. But a goal of this book is to find bridging concepts that enable this imagining to be done in a somewhat rigorous way, a way that has more of a chance of being true to the animals' lives.

I said above that some of what falls under the problematic umbrella of "subjective experience" is an expected consequence of animal evolution—the evolution of sensing and action, the formation of a point of view, and so on. But some philosophers will say that none of this takes us much distance toward an understanding of experience itself, because it does not grapple with the hardest problem in this area. That is the problem of explaining the *intrinsic qualities* of experience—the redness of red, the particular sound of a clarinet. Seeing the green of a forest in front of you has a feel to it—a "raw feel," as it is put—that is distinctive and very hard to explain in biological terms. The one-word term that has become standard—and notorious—for these features of experience is *qualia*. How do they fit into the world, and what role could they possibly play in any evolutionary process?

Some critics, like the philosopher Daniel Dennett, try to show that the whole idea of qualia is a mistake, an illusion. This critique seems literally mad to others, for whom nothing could be more undeniable than sensed colors and sounds. I think some of the problems that qualia pose are real; they can't be altogether wrangled from the scene. But I want to put them in their place, and make them seem less radioactive from a biological point of view.

When a person with normal vision looks at a tomato, and experiences its color, the biological goings-on in that person will have their own physical features and peculiarities; they will have an "intrinsic" nature. That will give the experience a particular feel for the person themself, the person who *is* this pattern of biological activity. A central challenge in this area is to say why brain processes of such-and-such a kind are the ones that give a person a feeling of red, as opposed to something else. That is indeed a challenge—a scientific one. But some views of what the scientific story ought to do here, some demands made of the materialist, become unreasonable. A scientific description can't encapsulate or *contain* an experience being described; knowing about an experience is different from *having* it, even if knowing about it might help you to imagine it. Some critiques of materialism seem to want a third-person description of a human or other animal to do something it could not, in principle, do; they want it to magically become first-person.

When people worry about qualia, they sometimes also set things up like this: the materialist is supposed to describe a collection of physical processes from the third person, and do so in a way that conjures up qualia. The reds and greens and sounds of cymbals are supposed to somehow *appear* in the system being described. This is entirely a mistake. The qualia are not extra things that need an explanation, somehow produced by the workings of the physical system. Instead they are part of what it is to *be* the system being described. Experience is the first-person point of view of a complex living system of a certain kind, not something conjured up by the workings of that system.

Qualia tend to overstep reasonable bounds in another way, too. The examples that philosophers fixate on become a model for all experience. If you fixate on the redness of red for long enough, you can end up thinking of experience itself as a procession of these colors and sounds. The paradigm case of experience becomes an encounter with a pure color. We might call these "Rothko experiences," after the color-field paintings of the American artist Mark Rothko. This label is not just supposed to be vivid, but also diagnostic. I think that Rothko's paintings feel the way they do because of their anomalous relation to the usual business of seeing. Human seeing usually includes various kinds of probing, and visual experience occurs against a background in which the subject is making sense of the spatial layout before them. But a color field seems neither within us nor without. We can indeed have this kind of experience, but it is not how seeing usually works. I think this disembodied character is part of the reason why Rothko's paintings are alluring and popular.

Taking this further, I will make use of some more ideas from Susan Hurley, who pushed particularly hard at these relationships before her very untimely death at the age of fifty-two. Hurley brought to philosophy a distinction, used in the psychology and neurobiology of vision, between "what" systems and "where" systems in our brains. "What" systems handle shapes and colors; "where" systems handle spatial arrangements. These two kinds of information are routed differently in the brain, to some extent, though the distinction is just a rough and ready one. Shape, for example, is a matter of *what*, but shape involves spatial arrangements between parts (*where*). In everyday seeing, the two are thoroughly mixed together.

These two aspects of vision differ in how they relate to action and feedback. In normal circumstances, your sense of where things are is continually modified by your own movement, and can often be cross-checked with touch. This aspect of vision is integrally related to a sense of *you versus the world*. You are one object, changing your location among other things that also move. Colors are part of how you keep track of this, making use of contrasts and the ways colors form shapes, but colors themselves are not usually tied to action in the same ways. You cannot commonly cross-check colors with touch, and in the case of a color field, where only the vaguest shapes are present, your brain's "where" system has very little to go on. Hurley thought that the sort of processing that a "where" system does for us is "basic to having a unified perspective or a point of view as a perceiver and agent, a sense of being a self present in the world, and hence to having a mind." Color-field experiences, in which this "where?" side of vision can get no purchase, are real experiences—I don't deny that—but they are not nearly as paradigmatic as some people think.

How did people's thinking in this area evolve to make the problem appear as it does? How did qualia become so central?

The ancestors of qualia were born and flourished in the seventeenth, eighteenth, and nineteenth centuries. These were the "simple ideas" and "impressions" of the empiricist philosophy of those days, especially as seen in John Locke, George Berkeley, David Hume, and J. S. Mill. Simple ideas and impressions were pure sensations, like patches of color and brief sounds. They were thought to populate the mind like a collection of mental atoms. In some philosophies, these were just about all there is, and even when that was not so, they dominated views of perception and experience. These pure sensations had a dual role for philosophy. They were part of an attempt to describe what minds contain and how they work, and were also pivotal to attempts to find new foundations for knowledge, sweeping away dogma in the process. If we reconstruct all knowledge as the tracking of patterns in sensations, a lot of obscure intellectual debris can be cast aside.

This picture of the mind stayed around, with minor changes, for a long time in English-speaking philosophy. In the early twentieth century, things called "sense data" played a dual role of the same kind. Philosophy today does not have "simple ideas" or "sense data," but their shape has lived on in qualia.

From around the end of the eighteenth century, a revolt arose against sensation-based views of mind and knowledge. Those views were criticized as rendering the mind entirely passive. Other faults were found as well, but passivity was often the heart of the objection. The German "idealist" project in philosophy rejected passive, atomistic views of experience, and tended to go to the other extreme, asserting the primacy of self-determining, autonomous consciousness. This area has seen a succession of overstated views on different sides.

In recent debates about sensing and action, similar stark oppositions arise. A view called *enactivism* tries, at least in some versions, to explain perception itself as a form of action: "seeing is a way of acting," and experiences are "things we do." These views make use of the feedback between action and sensing—the fact that what you do affects what you sense—and try to use this to bring sensing entirely onto the action side of the ledger. Once I say it like that, this view probably seems to be going too far, and that is exactly what I think. The aim of such views is to move as far as possible from a picture in which the perceiving mind is a mere screen, or a passive receptacle where qualia appear, but these views go so far that the receptive side of the mind—which is real—is denied altogether.

An ongoing feature of philosophy is its generation of wildly exaggerated theories. As the American philosopher John Dewey wryly noted, a philosophy student goes from being startled by the enormous differences between rival positions (everything changes; no, change is a mere illusion) to casually moving these ever-toppling pieces around the chessboard themself. This is a pathology of the field, though it is often the route by which new ideas appear—one stark and stylized picture is set against another. In the particular area of understanding experience, for some reason it seems hard to simply acknowledge *traffic* between living systems and the world, with sensing and acting coexisting in a two-way process. Philosophical attention seems to veer wildly from one side to the other.

Beyond the Senses

The topic of this chapter is experience and subjectivity, but so far it has largely been about *sensory* experience. I criticized some of the ways philosophers have thought about sensing, but another problem with much recent work in philosophy is the idea that sensing is not only an important part of experience, but just about all that goes on there. This is also something we need to move past, if we are to better understand how experience fits into the world.

Philosophers who think that all experience is sensory do accept that some feelings come from inside us. They use "sensing" to cover the detection of internal goings-on (such as hunger and fever), as well as of external things. We have both inward-pointing and outward-pointing senses. These philosophers also sometimes talk of "perception" rather than sensing, but those are very close. Here are two examples of the views I have in mind, from philosophers of different generations. Jesse Prinz, who I worked with in New York, is blunt: "All consciousness is perceptual." Fred Dretske, one of the philosophers who most influenced me when I was a student, who I worked with at Stanford at the end of his career, thought something similar, though he hedged the claim: maybe not all consciousness is like this, but "the clearest and most compelling" cases are found in "sensory experience and belief."

How likely does that seem to you—that the "clearest and

most compelling" cases of consciousness are sensory experience and belief? Surely, equally clear and compelling are emotions, willings, moods, and urges. Speaking for myself, I'd say that those are quite a bit *more* clear, as cases of conscious experience, than beliefs.

Perhaps emotions and moods are perceptions of hormonal activities, and other states of the body. The project here is to treat all kinds of experience as sensing or tracking or registering something. Your experience of being in a bad mood, in this view, is your detection of something going on inside you. An alternative to this view, rather obvious but neglected, is the idea that a mood is not a *presentation of* some fact or condition; it is just *the way things are* with you, at that time.

Here is another example. Consider energy level, especially fatigue. The best example is not physical fatigue, from muscles overexerted, but mental fatigue. Suppose you are driving along and your energy level shifts, heading toward fatigue. A heaviness and dullness sets in. This feels like something; it is part of experience. Does it seem likely that this is a presentation to you of some state of your body, like a fuel gauge showing a "low" reading? The alternative, again, is that your experience just includes, as if saturated by, this foggy, heavy way of being. There is a feelable difference between thought processes being slow and effortful, or agile and easy. Energy level is an aspect of your living activity, one that is felt.

Consider also *surges of resolve*. These, too, lead away from the view that experience is all about detection and perception. They push toward seeing experience as an aspect of living activity as a whole. Experience is not solely a matter of being *told* something. More of life is felt.

How much more? Here is a quote from the philosopher John Searle.

> Imagine that you wake from a dreamless sleep in a completely dark room. So far you have no coherent stream of thought and almost no perceptual stimulus. Save for the pressure of your body on the bed and the sense of the covers on top of your body, you are receiving no outside sensory stimuli. All the same there must be a difference in your brain between the state of minimal wakefulness you are now in and the state of unconsciousness you were in before. . . . This state of wakefulness is basal or background consciousness.

"Basal" in this context means fundamental or most basic—base level. Searle in this passage does seem to be describing something real. Its significance is less clear. One way of following this thread is to conclude that consciousness need have nothing going on *in* it; it is just a state of being. That is far from most views in psychology and philosophy today—most views approach consciousness as something like a way that information can be presented in the mind. If so, there has to be information that is being presented. Such a person might reply to Searle that in his waking-up scenario, there will always be the beginnings of an inner monologue, a registration of slight hunger, or something else of that kind. On the other hand, within neuroscience, in contrast to philosophy and psychology, some prominent figures, including Rodolfo Llinás, have sketched views of consciousness that are similar to the "state of being" view that Searle expresses here—views in which consciousness will usually reflect information coming in through the senses, but does not depend on it.

Perhaps there's another way of describing what happens when you wake and regain faint consciousness in a dark room. You once again feel *present*.

The idea of a sense of presence plays an uncertain role on the margins of recent discussions of experience. Sometimes it is just an evocative word attached to vague hopes. Presence is said to be the feeling of your being real and present in a scene. It's quite difficult to fill this idea out, but a good way to see what it might contribute is by looking at contrasts—at views that are entirely different. A few pages back I talked about the idea that all experience is a matter of perceiving things, registering what is going on. Some people who accept this also believe a view that is known as *transparency*. This is the idea that in experience we are never aware of our self, or of our experiencing, but only of things put before us. Experience is transparent and we see the world (including our bodies) *through* it. Related ideas are sometimes presented by people writing about meditation, claiming that meditation reveals the unexpected absence of the self. If this transparency view is right, then conscious experience always points toward something else, and consciousness itself is no more than this pointing or representation.

"Transparency" is an example of a self-erasing tendency in many views of experience. The idea of presence pushes back against this approach, against the idea of the subject as a mere medium or vessel. The self is not the topic, the focus, of most everyday experience, but neither does it vanish. At least for some of us, the feeling of presence in a scene is an important part of experience.

What *is* this feeling, though? Once presence is recognized, it is tempting to see it as an automatic feature of just being a living organism located in the world. In a primordial way, we can feel some things about what we biologically *are*. If so, a basic kind of experience comes "for free" in anything that is alive.

That is an appealing idea, but it is probably too simple. The feeling of presence has more behind it than that; it seems to depend on a lot of complicated processing that goes on continually within us, largely behind the scenes. This includes the ongoing monitoring of your body, and of how events there relate to what is going on around you. These background goings-on become evident especially when things go wrong.

The feeling of presence is related to a sense of ownership of your body, and that sense is subject to a wide range of pathologies and twists. A case to start with is the "rubber hand illusion." If you view a fake hand being stroked with a brush, exactly in time with strokes being applied to your real hand, which is out of sight, this can produce the illusion that the fake hand is yours and you are experiencing, in touch, the strokes you can see. The rubber hand illusion is also the tip of an iceberg. A wide range of illusions arise, sometimes through brain injury but inducible to some degree in experiments, in which a person feels themself to be somewhere other than where their body is. Just about every possible distortion is found. A complete "out-of-body experience" is one case, along with others in which you half-see and half-occupy a body that is projected as an image into your field of vision. In psychiatry, disturbance of a sense of presence, such as a "feeling of unreality," can be an important symptom of a looming problem.

If the sense of presence always arises from complicated behind-the-scenes processing, it is more than an automatic consequence of being alive in a living body. Perhaps, in reply, the sense of presence might exist in both simple and primordial forms and also in more complicated forms, which rely on internal monitoring and the like. A simple and inevitable feeling of living existence might be shaped into different forms in animals with different bodies and nervous systems. This idea is tempting, but I don't yet see a reason to believe it. Presence *feels* primordial, but that is not a reliable guide to what is going on.

Presence does not seem to be essential to conscious experience. A "feeling of unreality"—something people report that is different from presence—is still a feeling. But we are in the vicinity here of an important piece of what is needed to "bridge the gap" between the biological and the experiential. A sense of *thereness* contributes a lot to what seems special and distinctive about experience; it is an important part of subjectivity itself. As we learn about the activities in our brains and bodies that give us that sense of thereness, we go further toward fitting experience into the biological world.

Suppose a biological description is given of the workings of vision, for example. That story is told in terms of light, eyes, paths into the brain, and so on. It can often seem that this story leaves something out, and does not capture how vision feels. I think that part of the reason such a story seems incomplete is the fact that the experience of seeing usually includes, or accompanies, a feeling of presence. That feeling of presence—soft, elusive, almost entirely in the background—is part of why seeing feels the way it does.

People, whether philosophers, scientists, or anyone else who is reflective about these things, may have particular kinds of experience that they feel are indicative or telling about the nature of consciousness or subjectivity. Informal impressions of this kind are not trustworthy, but it is hard not to feel guided by them sometimes. For me, those indicative experiences are ones where there is a certain kind of balance in place between a feeling of my own presence and a taking in of what is going on around me. This state of mind is not self-absorbed, inward-looking, or introspective. Neither is it one where you might seem to disappear into transparency, left with just the scene itself. Instead, it involves a balance between my presence and the presence of surrounding things. This "balance" arises in some contexts of meditation. There is a scene, plus the feeling of being part of it. That feeling is a useful corrective given the way that theories in this area, especially in philosophy, tend to either overinflate the autonomous self or erase it. People perennially tend to amplify one side and elide the other. Instead, we can see through the errors of transparency without cutting off the world outside.

I said earlier that this chapter would have a mixture of reaching out and pushing back—reaching across the mental/physical gap, and pushing back against misconceptions that make the problem seem worse than it is. First there was some reaching out. Evolution has shaped animals not just into complicated collections of cells, but into centers of agency and subjectivity. The formation of these units is part of the origin of experience. There is a handful of different ideas and themes in this area, and I am not sure how much weight goes to each. One is the sheer tying-together of these new units, and the formation of an overall state of living activity in animals that is modulated by sensing. Another is the way that sensing and acting combine to yield an implicit—or explicit—sense of self versus other. In the case of human visual experience, the philosophers' favorite, all this comes together into a tracking of *what* confronts us, *where* things are, our sense of self-and-world, bodily ownership, and more. Given all this, it is not mysterious that seeing feels like something, rather than being a mere taking-in of information, as would occur in a camera. And though philosophers have spent a great deal of their energy thinking only about the sensory side of experience, especially those Rothko-like washes of color, more of life contributes to feeling than this. How much of it does—how many processes often treated as "unconscious" exert subtle influences on felt experience?

The story is far from complete at this stage, but this is a start.

Night Dive

Tom Davis, the soft coral diver from Chapter 3, and I were the only people who walked, about an hour after sunset, down from the roadside into the still water of the bay. We made our way through the shallows and the first puzzles of that site—flashes of red, unknown movements in the weed. As we went deeper, I was struck by the night sea's loneliness. If you turn off your diving lights for a moment, you are immediately pressed by inky blackness. Nocturnal animals on land often have a fair bit of light. On a dark night in the sea, once you are thirty feet down, almost all the light has gone. The animals at home there are immersed in scents, tastes, and touches.

Tom was surveying rare fish—tropical visitors to the bay that make their way under ledges at night. He found some, but the ledges were also full of sleeping fish familiar from the daytime, big wrasses and morwongs huddled together. It was hard to see the small visitors among the bunk-bedded locals.

Near the end of the previous chapter, I described a small diorama with a decorator crab. In the daytime the decorator crabs one encounters are covered mostly in sponge. At night, along with those, you find decorators clothed in soft corals. It is as if they are members of a secret society, a covert crustacean order, distinguishable by their garments and only seen at night. A decorator of this kind marched over the reef. Along its body, reaching fingers of soft corals extended into the sea. About half the coral hands were wide open and half were clenched.

As big fish slept under ledges and the decorator marched, a pair of hermit crabs in shells fussed with each other face-to-face. One had plastered its shell with many anemones; the other had just a few. The anemones, made mobile by the walking crabs, encountered surprises. Their tentacles were outstretched, touching the other crab's shell sometimes and reacting, pulling back. I didn't see two anemones themselves touch across their crab-drawn carriages, but they probably did at some stage. Many creatures afoot in a dark expanse.

6

THE OCTOPUS

Rampage

A while ago I swam out on another dive to one of the soft coral sites described in Chapter 3, during the stillness at the change of tide. Usually that pause sees a degree of tranquility. Some rather strange-looking animals amble about, but minding their own business—a quiet coexistence of eccentrics. This time an octopus was on the move; I arrived to see her standing and setting out. This was a medium-sized octopus, with a body not much bigger than a softball. She was not large for the species, but extremely active.

I began to follow her. I am not certain that she was a "she." It is hard to sex octopuses, but males often move in a way that gives some protection to a particular one of their arms. This, in most species, is the third right arm. An octopus's eight arms circle its mouth, but if you look at the animal from the front, the arms are arrayed in a way that presents two in the center—the first arms, left and right. Outside them are the left and right second arms, and so on back to the fourth pair. The underside of the third right arm in males has a specialized duct used in mating, and males tend not to expose that arm as much as other arms.

This octopus's right third arm was moving all over the place with the rest, so I will treat her as a "she." She repeatedly went into an enveloping eight-armed bear hug around a bit of coral or a sponge, and all sorts of creatures would exit in response. It was interesting to see who was worried and who was not. Some of the most alarmed were small and previously well-hidden octopuses, who would scurry away watchfully, or flee with a jet. The decorator crabs, in their protective and deceptive clothing of sponge, looked completely safe. On the several occasions an octopus arm touched a decorator crab's cloak, the octopus moved on without reacting.

Surprisingly, seahorses also seemed safe. As the octopus came barging in, seahorses would float up like porcelain birds, birds flying slowly by flapping fins and sideburns. Again, occasionally an octopus would touch a seahorse in this process, and neither seemed to much note the interaction.

Flatfish, buried in the sand, were definitely worried. They scampered up and fled. But the animals who seemed most nervous were crabs of other kinds, those not protected by a coat of sponge or coral. They would come out very fast and sometimes the octopus went after them vigorously. One crab swam straight up toward me and the octopus jetted after it. The crab was in my gear somewhere and briefly the octopus was, too. The octopus extricated herself, and I guess the crab escaped, as the hunt continued.

A bull in a china shop; an octopus in a soft coral bed. At the end, she abruptly stopped, began burrowing down, and wedged her body under a clump of shells and other debris. I don't know if she caught anything in the entire nearly hour-long hunt, but she certainly raised hell along the way.

This wasn't a particularly big octopus, as I said, but she dwarfed much of what was around her that day. And the octopus was far more aggressive and active, scattering other animals before her. On a modest scale, it was reminiscent of the time when cephalopods ruled the seas.

When Cephalopods Ruled

Octopuses are molluscs, part of the same large group of animals as snails, oysters, and clams. They are in a particular group of molluscs, the cephalopods, along with cuttlefish, squid, and a few other curiosities.

Molluscs date perhaps from the Ediacaran, certainly from the Cambrian. The mollusc body itself is soft, with no bones or external skeleton. But from the Cambrian onward, many molluscs have had a substantial defense, a hard mineralized shell.

This is an unlikely body, it seems, to become active and evolve complex behavior. But soon after the Cambrian, some early cephalopods lifted off from the seafloor and ventured into the open water—or probably, in some cases, adopted a hovering crawl. Their shells enabled buoyancy, and a cluster of tentacles around the mouth gave them a new behavioral tool.

In the Ordovician, just after the Cambrian, some of these cephalopods grew large—up to eighteen feet long, with a head, a cluster of arms, and a long conical shell behind. They were the largest predators of that time, supplanting arthropods in that role. These active shelled cephalopods flourished in the seas for several hundred million years, but now almost all are extinct.

They left just one small representative of that experiment, the nautilus, still found today in the Pacific, ascending and descending in a quiet daily rhythm.

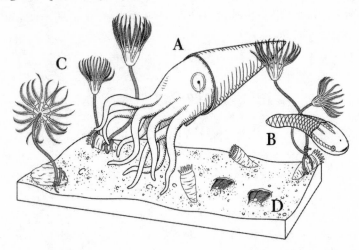

A scene from the Ordovician. Animals include **A:** a large straight-shelled cephalopod, *Orthoceras*; **B:** an *Astraspid*, an armored fish of a kind encountered in Chapter 7; **C:** a sea lily, related to starfish (*Glyptocrinus*); and **D:** trilobites.

As the giant shelled forms were receding, during the time of the dinosaurs, another group of cephalopods was taking a different approach to a life more ambitious than that of the typical mollusc. The *coleoids* are a group of cephalopods that had diverged from others around the heyday of the large shelled forms. Through the next era, the Mesozoic, they made one of those moves that turned out to have a surprising range of downstream consequences: they internalized their hard shell, abandoning its protection but retaining a version of it for buoyancy. This group of animals split several times, and in some cases began to abandon the shell altogether. Just one group went all the way in this process, ending up one day with no shell, inside or out, and almost no hard parts at all. The result, something like 100 million years ago, was the octopus.

Octopuses have the largest invertebrate nervous systems and are very behaviorally complex. They are complex in a particular way—exploratory, manipulative, often interested in novelty. They have several distinct ways of moving, including jet propulsion and various kinds of ambling crawl. Octopuses are so active and complex, and so far from us in genealogical terms, that they are a particularly notable evolutionary experiment. I went through the basic shape of the tree of animals in Chapter 4. A special place in that tree is occupied by the last common ancestor of bilaterian (bilaterally symmetrical) animals. From this unknown animal, perhaps 600 million years ago, two tiny genealogical shoots emerged that eventually became huge branches of animals. One of these branches is ours. The other includes molluscs, arthropods, and most other familar invertebrates. Among all the animals coming out of these old branching events, just three groups produced species with large nervous systems and complex behaviors. Those are the outcropping of molluscs called cephalopods; arthropods; and vertebrates: the rampaging octopus, the crabs sent scattering, and the gear-festooned human following.

Debate continues about how much complexity was present in the common ancestor of all these animals. For many years, people have had in mind something like a flatworm. But recent work has raised the possibility that the animal was more complex than had been supposed. Some biologists looking closely at insects have suggested that subtle similarities of design between ourselves and arthropods show that a small "executive brain" was present all the way back in the early bilaterians—this has been argued by Nick Strausfeld and Gabriella Wolff, and would be a change of view. An executive brain in their sense is not something extravagant—it would be a pretty small cluster of controlling machinery in the front of the animal—but it's more than people had supposed might be present. As Strausfeld acknowledges, though, if an executive brain was present at that early stage, molluscs then threw it away. Molluscs gained shelter in their shells and had little need for a nervous system built for an active life. Cephalopods rebuilt a complex brain much later, and did so on a different design. That brain does use much of the same chemical toolkit seen in other brains. In 2018, some octopuses were given the drug MDMA, or ecstasy, to see how they would respond. Surprisingly (at least to me), they seemed to become more friendly and gregarious, a bit like their human counterparts. But even if the main items of this chemical kit are conserved between other animals and octopuses, the layout of their brain remains profoundly different.

This distance and independence of their large-brained evolutionary path is one reason I find octopuses so interesting. Octopuses also lead to reflection on different ways of being an animal, and the different kinds of experience that might be associated with different animal bodies.

Consider the contrast between an octopus and the banded shrimp I visited in Chapter 4. Arthropods and cephalopods were the first large predators in the sea, dominating different periods in turn. Compare their relatives now. The arthropod way of being, especially in insects and crustaceans and the like, is based very much on hard parts, parts that shape and scaffold the animal's actions. A great deal of effective behavior is available to such a body, but the body is also, in a sense, rather closed, or restricted. It is a body of springs and spatulas, of claws and legs and clubs. Earlier I compared this body to a Swiss Army knife;

that is a good analogy because while a Swiss Army knife can be used to do many things, it also embodies restrictions due to its specialization and fixity. Similarly, the arthropod body is full of hard parts that do just what they do, and no more. An octopus, in contrast, can grasp and manipulate just about anything—as well as doubling the length of its arms or flattening itself like a pancake. It has an openness of body and action.

Lines of Control

The cephalopod body, flexible and muscular, has great potential for action but brings with it special demands. In an octopus arm, the "degrees of freedom" are almost countless. It is hard to organize this body, to make actions work, but a great deal can be done *if* you can get it to work.

Perhaps for these reasons, octopuses embody a design very different from our own. The octopus nervous system is decentralized; about two-thirds of the neurons are not in the brain (itself a vaguely defined region), but in the arms, especially in the upper arms. These act not just as outlying sensors and relay systems feeding the central brain; there is an apparent delegation of the control of some motions to the arms themselves.

This can be seen both as a response to a difficulty and as an opportunity. Decades ago, Roger Hanlon and John Messenger suggested that the large and decentralized octopus nervous system might exist because of the difficulty of control of that body. This is also an opportunity, though. Once you are growing eight free-moving arms in a nest around the mouth, why not imbue them with an array of sensors, and allow them to go their own way to some extent?

In many situations, octopuses apparently behave as a whole, but it has long been unclear exactly what the relationship might be between the central controlling brain and those neurons in the arms themselves. It was suspected in early years of research that central control over the arms is quite limited, and an octopus might even have little knowledge of where its arms are at any given time. This was suspected because of the animals' difficulties with some artificial laboratory tasks, and also by the fact that the connections between arms and brain, within the nervous system, were rather slim.

The lab that has looked most closely at these questions since then is Benny Hochner's in Jerusalem. A clever experiment done there showed that an octopus can use vision to send a single arm along a novel path, out of the water and back again, to reach food. This seems to show considerable central control. The report of the experiment, by Tamar Gutnick and her collaborators, also notes that when octopuses do the task well, there seems to be some local exploration by the arm as it goes—though this is just an impression people have when watching them. The same lab has reported neurobiological work suggesting that octopuses do not have a map of their own body within the brain, as we do. This suggests that if an octopus does have some sense of how its body is arranged at a given moment, it achieves this quite differently from us.

A decentralization of the body's controllers is seen not only in the octopus. Forms of it are found in many animals, including ourselves, and there are reasons why this might be expected to some extent. The features of animals that are responsible for sensing and action, the features whose history I have been charting, are complicated and require many parts. Organized arrays of cells are needed for seeing, other arrays of cells are needed to produce movement, and nervous systems with still more cells are needed to coordinate it all. Once you have all this machinery, you have some new evolutionary options. You can separate out some pathways, create local lines of control. It becomes a choice, in an evolutionary sense, whether you create separate streams that control action, or integrate everything into a single stream. You can also do a bit of both; you can have largely separate tracks with some cross-talk between them.

This is important to the themes of this book for a number of reasons. I have been using the idea of "point of view" as part of what makes sense of the idea of subjectivity. *Point* of view has always been metaphorical, but it suggests a lot of integration. In fact, many animals are only partially integrated. When you separate out the streams, you tend to gain some things—often you may gain speed—and lose some things. If sensory streams are separated from each other, you will lose the ability to combine different kinds of information that might be useful when considered together, like the many premises of an argument or many pieces of a puzzle. If you allow different parts of yourself to perform their own actions, you risk a situation where your active parts want to do contradictory things. In the extreme, you risk fragmenting entirely, into sub-agents who see their own scenes and make their own decisions. That certainly looks like a bad idea, but we should not assume that everything follows a shape with a central CEO and a lot of underlings.

An important case that is close to home is *lateralization* in the brains of animals like us. We are, as discussed earlier, bilaterally symmetrical animals with left and right sides. This means that many parts of our bodies are paired, including legs, lungs, and much, though not all, of our brain. The two sides of the upper parts of our brains are connected by a thick channel of fibers. In vertebrates, there is also a notable crossing-over of lines. Events seen in the right visual field are processed by the left side of the brain, and vice versa. In invertebrates that are bilaterally symmetrical, brain parts are often paired to some extent, but not crossed over in this way. In an octopus, for example, there is a big "optic lobe" behind each eye, and these deal with the eye on their own side.

This pairing of parts in animal brains has resulted in some surprising divisions of labor between the two sides. For example, various vertebrate animals have a left eye preference for handling social interactions with other animals of the same species, and a right eye preference for dealing with food. Again, if the left *eye* is preferred for social interactions, that means the right side of the *brain* is used, and so on. In cuttlefish, which are cephalopods like octopuses, the right eye is preferred for feeding and the left for dealing with predators. As cephalopods don't have crossed wires, left eye means left brain. They make a similar division of roles even though their brains are different in design.

The radical human incarnation of this, one not arising from a natural event (or almost never) but instead as a surgical outcome, is "split-brain" patients. Some people who suffer severe epilepsy have the bridge between the two halves of their brains cut to prevent seizures spreading from one side to the other, as a seizure occurring in half the brain is better than a seizure in all of it. People who have undergone this process seem in some circumstances to have two minds in one skull. In other circumstances, they behave quite normally. I'll take a close look at split brains later; they provide clues that might help us unravel the puzzles of octopus experience. But before that, let's do some octopus watching.

Octopus Watching

A setting where the perplexities and charms of octopus behavior are especially evident is a pair of field sites in Australia where I have watched these animals for about a decade now. The first site was discovered by chance in 2008 by Matt Lawrence, while he was doing exploratory dives in a large bay. This bay is not the site of the scenes I have been describing in earlier chapters; that is Nelson Bay. The octopus sites are about a six-hour drive south, on the same coast. Wandering on scuba across a sandy plain populated by scallops, Matt came across a small area where thousands of empty scallop shells were piled up and about a dozen octopuses were living. We call that site "Octopolis."

Octopuses have generally seemed very solitary animals, and many species probably are, but this site showed that octopuses can, in some circumstances, live in rather close quarters. The largest number of animals we have seen at that site is sixteen. They live there in both sexes and in many sizes. In a sense, they are also present in "all ages," but octopuses have surprisingly short life spans of just one to two years. Many generations have passed during the time we have been watching. A large octopus down there will have a body about the size of a football and arms several feet long. Others are no bigger than a matchbox, and these can only differ in age by a bit over a year.

The origin of Octopolis is unknown, but we have some ideas. The site has unlimited food for an octopus, but it also has many predators—sharks, seals, dolphins, and aggressive schools of fish. The sand is fine and silty, and it is hard to build a den that will provide much safety. The site exemplifies the dilemma of octopus life, a life spent as both prey and predator. In this setting, we think that a single human-made object was probably dropped from a boat some time ago. That object, just a foot or so long and probably made of metal, is largely buried now. But it may have seeded the site like a crystal. An octopus, or perhaps two, could build a good den adjacent to it, and bring in scallops to eat. The octopus(es) ate the scallops and left the shells. As shells accumulated, these began to provide a better building material than the silty sand, and more octopuses became able to build a den there and live safely. These brought in more scallops and left more shells, and soon a process of "positive feedback" was underway, where each new foraging octopus inadvertently created living opportunities for others.

Whether or not this is exactly what happened to get things started, it is clear that the octopuses bring in scallops and eat them and leave shells, and the shells make it possible to build high-quality dens for themselves and others. These dens are in some cases vertical shafts like mines, more than fifty centimeters deep, with straight shell-lined sides. They are safe and secure.

The dens are secure in relation to non-octopuses, at least; a feature of this site is a lot of territorial wrangling and eviction. One octopus will dive into a den and haul out another. The two will wrestle, and the loser will jet away. The winner sometimes occupies the den, but sometimes just goes back to where he or she came from.

This site is an important one for attempts to understand these animals, as octopuses here have had to work out how to deal with each other, how to handle the continual presence of others of their kind. The most complicated things in most animals' environments tend to be other animals, including—and especially—other animals of the same species. Octopuses often don't have much to do with each other, it seems, but in Octopolis, they are well and truly surrounded by others. Our octopuses have to navigate this complexity. They do so with a certain amount of aggression, but the hostility is often rather mild. They also show a number of behaviors that seem to be aimed at monitoring who is around and who is who, perhaps especially in relation to sex. An octopus will come into the site, ambling or slowly jetting, and as it passes different animals there will be a series of arm-raises, slaps, and touches. Some of these have the appearance of, or quickly turn into, a sort of boxing or sparring, with a rapid flurry of probing arms. But many other interactions are not like that. Often there is a reach, either one-sided or two-sided, then just a brief touch as the animal passes. The animal being passed will settle back into its place or its den.

Occasionally these interactions do escalate into fights, but my impression (very informal—not yet based on numbers) is that most real fights start differently. The big fights happen when an octopus marches or creeps in toward the site from the outside, and another comes out to meet him and they battle in a many-armed frenzy.

We have seen countless fights at this site now, but in all the ones I've seen, there has never been a death or obviously serious injury. I have come to think it is quite hard to hurt another octopus with their available weapons, unless there is a big size difference between the animals. Deaths have certainly been seen in octopus-on-octopus conflict elsewhere, apparently from strangulation. I suspect those cases had substantial size differences. Otherwise, if the animals are of similar size, a lot of octopus conflict looks like (as Miranda Mowbray said after seeing one of my videos) a giant pillow fight, between pillows.

Males in some cases appear to actively exclude other males, and they seem able to work out from a touch which individuals they do not want to exclude. A touch may indicate sex, but only rather fallibly—sometimes our octopuses seem to get the sex of another individual wrong. Males have snuck in under the noses of others and settled next to females, and mating attempts have sometimes dissolved because, I suspect, the sex of an animal was initially misidentified. All this often takes place in an ongoing ruckus of jostling and turf-wrangling.

We are not sure whether some of the behaviors we are seeing here are genuinely new, or at least uncommon for the species. This might be a situation where animals of a mostly solitary species have come to live in unusual densities and are learning, individually, how to get along. Alternatively, more sociality might exist in octopus lives, in this species and some others, than has been recognized.

That picture of Octopolis, with all its uncertainties, has been in place for a few years now, and I described it in my earlier book *Other Minds*. An event that helps considerably with all these questions occurred in 2017. Two other divers, Marty Hing and Kylie Brown, exploring the same general area as Octopolis, discovered a second site, one with similar numbers of octopuses and similar activities afoot. But at this second site, now christened "Octlantis," there's no apparent role for a human-made object in getting it all started; this is a wholly "natural" site.

Again in a setting with much food, much danger, and few den options, a couple of isolated rocks poking up from the seafloor seem to have enabled a feedback process to get started. Octopuses bring in scallops and leave the shells. The dens at this site are in some cases nestled up against the rocks and in other cases built directly into the shell-strewn seafloor. Octopolis was not a unique event, dependent on an inadvertent human act. Octlantis shows that it can all happen again. The numbers at Octlantis are similar to the numbers at Octopolis; the most I have seen there is fourteen, spread across three subregions of the site, but all in a pretty small area.

Both these sites generate more behavior than you usually see with octopuses—more interaction, more activity. We don't do experiments there; we just watch the octopuses do whatever they want to do. But the years of observation—some of it organized and planned, some very informal—have given us a growing sense of their range of behaviors. As I watch, I am always wondering about those relationships between central control and the brainy arms.

A lot of what we see is coordinated, whole-body behavior. The animals shift between different kinds of motion. In jet propulsion, the arms are brought together and the animal becomes a slender missile. When crawling, the arms go everywhere at once. Some behaviors are packaged together into what appear to be social displays. An aggressive animal will often stand very tall with arms spread, and with the mantle (the large rear part of his body) pointing straight upward. This is combined with intense dark colors—an octopus can change its entire color in less than a second. The combination makes the animal look as large as possible, and genuinely ominous. We call this the "Nosferatu display."

A particularly intriguing behavior is throwing. Octopuses sometimes gather material in their arms, and then either carry and release it, or sometimes throw it out in a concerted, occasionally spectacular, way. The arms are used to first collect shells, seaweed, and silt (sometimes just one of these payloads, sometimes several). As this material is held, the animal's jet propulsion device is brought under the arm web, and a sudden jet of water propels everything as it is released from the arms. These debris-throws can go for several body lengths. Often the material thrown hits another octopus.

My collaborator David Scheel was the first person to note a possible social role for this behavior. Are the octopuses aiming these throws at others? Is this another form of the mild aggression often seen at the site? It is very hard to tell, as this requires working out what the octopus *intends* to do. That is difficult enough in the case of animals very close to ourselves, and extremely so with an octopus. David and I have spent a lot of time trying to work out how to interpret octopus intentions, an activity fraught with both scientific and philosophical puzzles. Here is how things currently seem to me.

Most debris throws are probably part of den-building and den-cleaning activities. Octopuses spend a lot of time clearing rubbish that accumulates in their dens, and throws are part of this. If an octopus is doing this while paying some attention to another nearby octopus, as they often are, then it can be expected that some throws would inadvertently result in hits. Females seem to throw more than males—that is intriguing, but females also tend to build and maintain better dens than males. This makes sense, as females eventually have to brood eggs.

I think that some throws probably do have a social role, though. Females quite often throw at males who are pestering them. On one occasion, a video shows a female octopus throwing debris repeatedly at one particular male over a period of a few hours. About half of these throws hit him, and others missed only because the male ducked or was belowdecks. Toward the end, the male who had been on the receiving end seemed to be getting used to these assaults; he began to duck quite early as the thrower began loading up, and the final broadsides went (mostly) over his head.

Throws with a social role make sense as a modification of two more common behaviors. One of these behaviors is using the jet to clear debris from the opening of a den, and the other is targeting the jet directly at octopuses and other animals. If you are hanging out with an octopus near its den while diving and you start to bother it—perhaps by interfering, perhaps just by being too long in its space—you will often feel a brisk jet of water propelled toward you. With those behaviors on board already, one can see a path where an octopus throws debris incidentally, hits another octopus by chance, and notices the effects. The effects are often considerable; if the throw is big, a pestering male will back off, startled. Females also throw and hit other nearby females, perhaps as an instance of the low-level aggression I mentioned earlier. I am not certain of any of this; with octopuses, it's always hard to tell what is going on.

The throws are interesting both for their possible social role and also because they are coordinated, centrally organized behaviors. Like jet propulsion, they require a particular arrangement of arms to hold the material in place. The building behaviors we see at our sites are also interesting in relation to these questions about arms and brains. Octopolis as a whole is not a deliberately constructed city, though the individual dens that animals live in are deliberately built, and tended quite carefully in some cases. The fact that bringing in scallops to eat creates opportunities for more octopuses to live safely at the site is, we assume, something the octopuses are oblivious to. Individually, they are rather good engineers, but as far as we can tell, there is no cooperation in building, even by pairs.

I saw an interesting use of a found object. A small octopus was staring at one of our unmanned cameras on a tripod for a while from its den, and then went off camera and returned with a piece of dead sponge. This it arranged on the top of the den, as something between a roof and a helmet, and huddled beneath it, looking out. The sponge was the best possible material for the task. It was the right size, rigid, and light. I am not sure that the little octopus was bothered by the camera, and wanted a barrier against its presence, but it did look that way.

All these behaviors—quick, coordinated across the body, and visually guided—suggest there is a good deal of top-down control. These octopuses can see well, too; they are not chronically nearsighted, as has been claimed. I have noted cases where octopuses saw other incoming octopuses from a considerable distance, and reacted in ways quite different from what we see when the incoming animal is a stingray or some other intruder. Octopuses watching something intently often also do a head-bob motion, up and down and a bit to the side. This, as Jennifer Mather has noted, seems to be an attempt to improve perception of the distance of an object, its depth in the scene, by changing the angle of view. Octopuses mostly see objects with just one eye, given the shape of their heads, and depth perception would otherwise be difficult. This is active information seeking in vision, and also something that requires handling subtle relations between what is called *reafference* (sensory changes caused by your own actions) and *exafference* (sensory changes caused by something going on outside you).

Octopuses show all these well-organized behaviors, but there's also another side to their actions: ongoing exploration by the arms. A crawling octopus that's not in a hurry often allows its arms to roam a bit in many directions. When an octopus is sitting fairly quietly, a few arms often range farther, looking like little eels with their delicate, questioning tips. I see this much more at other places—including Nelson Bay, site of the rampage at the start of this chapter—than I do at Octopolis and Octlantis. I suspect that this is because at these other sites, one usually sees octopuses on their own, not having to deal with other octopuses. Then they are more relaxed. I think there is often a kind of heightened attention in the octopuses at Octopolis and Octlantis, due to the social complexity of their surroundings, and ever-present questions of sex.

A picture suggested by all this is one in which the octopus body is subject to a kind of mixed control. The body can be partially commanded and steered around by the central brain, but that body has parts that engage in their own ongoing exploration, reacting individually to their surroundings. Centrally coordinated actions can pass over to the exploratory tendencies of the arms. Watching octopuses sometimes results in a series of gestalt shifts, between seeing the animal as a whole whose each arm is a tool, and seeing an arm wander about apparently in response to what it is sensing itself.

A famous neurological patient, with the apt name of Ian Waterman, lost all proprioception—that is, lost all internal sense of his own body's arrangement, such as where his arms and legs are—due to an infection when he was nineteen. He has had to learn to use vision to keep track of his own body. This was extremely disruptive, and recovering the ability to move and act coherently was difficult, though he prevailed. A few people have made a comparison to octopuses—as Fred Keijzer said to me, perhaps an octopus is a "natural Waterman." These interpretations draw on the work suggesting that octopuses do not have an internal map of their own bodies. But if octopuses are Watermen, they are handling their situation in a rather effortless way. Of course, this would to some extent be expected if the Waterman state was natural to them, and how they have always been. But if it is true that an octopus can only organize its arms for a complex motion if it can see where they are, that would be surprising. Octopuses do look immediately cohesive when they want to act quickly.

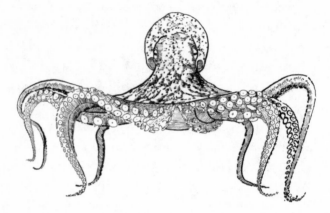

People often now talk about octopuses as "smart," and in some ways they are. But that is not the term that comes readily to my mind. Octopuses are behaviorally complex animals, and I think they are also sensitive animals; I think they experience their lives in a rich way. The word "smart" points toward a particular way of being, however. It suggests that we interpret their behavioral complexity in a rather intellectualized manner. Octopuses are exploratory animals who direct the complexity of their bodies on whatever confronts them. They fiddle about and try things and turn the problem over and over—physically, not mentally. Octopuses have an extraordinary sensorium and an anarchic bodily embrace of novelty, but they are not, for the most part, ruminative and "clever" sorts of animals.

They do have a little of this clever side—some famous incidents where octopuses have escaped mysteriously from aquarium

tanks might involve something close to planning, and their em-
ployment of objects like shells and coconuts for protection is a
kind of tool use. This use of objects seems to have an improvised,
opportunistic look about it in some cases, including the case of
the octopus at Octopolis who retrieved a sponge to hide from
our camera. Those behaviors suggest a sort of mental as well
as physical exploration. This appearance might be misleading—
perhaps all the uses of shells, coconuts, sponges, and the like
are well-established behaviors that evolution has shaped as a response to predators. But here, along with one other context, is
where I'd be looking for signs of "cleverness."

The other context is more social. Octopuses show a surprising awareness of what other agents, including people, are up to.
They often make their move to escape when you are not looking at them. The same sort of thing is true with cuttlefish. Bret
Grasse, who manages octopuses and other cephalopods at the
Woods Hole Marine Biological Laboratory in Massachusetts, has
spent more time around these animals than just about anyone.
Bret has the impression that they are often very aware of what
he is up to. They squirt water at him sometimes, but wait until he
is not looking. Once Bret was squirted and he turned to see a
bunch of innocent cuttlefish near the bottom of their tank. He
then used his phone's camera to watch them while his back was
turned. Several came up to the surface and squirted him again.

Something else I like about octopuses—not something related to smartness, just something good—is the fact that individuals of the same species differ so much, even when performing
fairly basic behaviors. They show many differences in personal
style, for want of a better word. Over near the rampage site on
another occasion, I came across a large octopus in his den. I

didn't try to disturb him, but he hauled himself out as I watched, and we set off across the landscape. He kicked one octopus out of its den and mated with another. All the time during this, he moved in an unusual, stylized-looking way, producing flattened blade-like shapes in his arms, winding his arms over his head and backward for no apparent reason, coiling an arm into a wheel. I had not seen an octopus handle its arms like this before, and there seemed no particular reason for it. It just seemed an eccentricity, a quirk, like the many individual quirks in den building. Everything he did was writ large.

Octopus and Shark

Arthropods and cephalopods were the first large predators in the sea, pillaging their way through different periods in turn. Octopuses themselves have never dominated their environments. They evolved many years after the first cephalopod golden age, and right from the start they were accompanied by some formidable competitors—fish—who have never been absent from their lives. Octopuses never had the field to themselves, as the rampager at the opening of this chapter appeared to.

At the end of a daytime dive up in Nelson Bay with the soft corals, heading in to shore and low on air, I saw a large, impressive octopus sprawled out on a flat area with rocks and weeds scattered around. I took a few photos and then realized the octopus had some arms active below its body. I thought it might be a mating, but after watching for a while I found that although another octopus was indeed beside and underneath this one in a crevice, the scene was not a mating but a lengthy tug-of-war. The octopuses were wrestling over a good-sized fish. It was a slow trial of strength, a contest that went on for over ten minutes. In other ways, the scene was reminiscent of a mating pair, with a female deeper in a den and a male hanging around outside—that is a common setup. Does it seem unlikely that you and your mate would wrestle for ten solid minutes over a single fish? Not with octopuses. I do suspect, given various details, that the octopus I had seen first on top of the den was male and the other was female, but in this case I will tell the story neutrally, with Octopus 1 and Octopus 2.

Octopus 1 eventually prevailed with the fish, which disappeared beneath its body. It went to an adjacent spot and sat there. Octopus 2 sat in the den.

After a while, a wobbegong shark came cruising in. These are sharks with thick, broad-beamed bodies and a camouflage pattern like an old bomber plane, in olive-green, brown, and gray. They tend to hug the bottom, often waiting in ambush for unwary prey but occasionally swimming around. They are not usually dangerous to humans unless you bother them, in which case they can give a significant bite—and a dangerous one, as they tend to bite and then hang on. This individual was not close to full-grown, as it was just three feet long or so. The largest are triple that size.

Octopus 1 looked a little agitated, and retreated, watchfully. It maintained a distance as the shark wandered about. I could not see Octopus 2, as that one was deep in its den. Suddenly attacking, the shark then plunged headfirst into the crevice after the second octopus. With body and tail near vertical, thrashing violently, it began wrestling its body into the den. Despite the intensity of the attack, it seemed to fail. After a while, the shark slowed its thrashing and stopped, and then sat back a little, still very close.

Octopus 1 had not moved off far or hidden—in fact it edged back closer during all this. I came around to see what had happened with Octopus 2, and saw that it was injured and very still. Octopus 2 had not inked at the shark, surprisingly to me. It just stayed low and held on. Soon the shark made another attack of the same kind, but it failed again.

The shark then gave up, and turned its attention toward Octopus 1. Again to my surprise, Octopus 1 seemed slow to react. It backed away, without urgency. Then the shark lunged forward, and it became apparent why the octopus had shown so little concern. With a jet, it effortlessly kept a safe distance. The octopus seemed to know it could escape once there was no element of surprise.

The situation now appeared to be a standoff. The shark moved behind a rock. The position it chose was interesting, as the octopuses probably could not see it. The shark's head was oriented toward the octopuses, behind the rock. Watching from above, I wondered if it was a rather sophisticated attempt at an ambush, with the shark cognizant of who could see who. But if it was such an attempt, no octopus was tempted. The shark gave up this position and wandered back into the weed.

Octopus 2 had not moved at all. It had what looked like a significant injury, but was definitely alive, still breathing. As octopuses had for millennia, despite sharing the seas with sharks, it survived.

Integration and Experience

Ten years of following octopuses around and watching them—especially at Octopolis and Octlantis, where so much behavior is seen—have left me with no real doubt that octopuses experience their lives, that they are conscious, in a broad sense of that term. This view is not just a response to their sheer complexity, and those active eyes with large brains behind them. Even theories in this area that attribute a lot of animal behavior to *un*conscious processes would have reason to put octopuses on the "yes" side of the line. Octopuses show an attentive engagement with novelty, including novel behaviors by humans, and much of what they do is not at all routine. They seem to undergo moods—stressed, inquisitive, playful. At Octopolis, we've seen a large male try to keep an eye simultaneously on several other octopuses, sometimes seeming to hesitate about who to chase and who to ignore. The octopus seems fairly clearly to be a conscious invertebrate. This is the clearest case, or perhaps one of two, along with Elwood's hermit crabs. That shows us something about the history of consciousness, given the shape of the genealogical tree. It shows that either consciousness has at least two or three distinct origins—one for us, one for octopuses, one for crabs (and perhaps more)—or, if there was a single origin, it was deep in time and took a very simple form.

The octopus also raises puzzles. I said in the previous chapter that part of the explanation for the evolution of experience lies in the origin of a new kind of *self*, a self tied together in ways that give it a point of view, ways that make it into a subject. A good deal of this can be described as a kind of integration of the animal. Integration has been a theme in quite a lot of recent thinking about materialism and consciousness. Sometimes it is seen as *the* crucial thing in explaining how a physical system can have experiences. In the octopus, though, we see an animal that is very complicated but less integrated. An octopus is still in many ways a whole, a center of action and sensing, but one organized in an unusual way. All these questions are made difficult by on-going uncertainty about what the animals can do, and how they are set up inside, but let's assume, for the sake of exploration, a view sketched earlier in this chapter. Octopus behavior in this view arises from a mixture of central and peripheral control. What might this feel like from the inside?

A first option is that the unusual design of the octopus does not make much difference. Although a zoomed-in view of the animal reveals a less integrated design, perhaps that is not very significant for the whole. An octopus often behaves in a very unified way. However, this does not settle the matter, as unity in behavior can arise by several different routes. Consider colonies of army ants, honeybees, and other tightly organized social groups—sometimes called *superorganisms*. In some respects they, too, act as wholes, but beneath the colonial unit is a collection of individual agents, each of whom senses and acts. Those colonies remind us that teamwork, undertaken by a number of different individuals, can be powerful in giving rise to unified behavior.

We should then at least consider the possibility that an octopus is a being with multiple selves. There is a primary or most complex self—the central brain—but also eight smaller ones. These smaller ones might not be sentient or conscious, but the general shape of the situation would be: 1+8.

Yet another possibility, a third option, is not just 1, or 1+8, but 1+1. The networks of nerve cells in the arms of an octopus are not only connected to the central brain, but connected "sideways" to each other, at the top of the arms. A few people have raised the possibility that the nervous systems in the arms are connected well enough among themselves that they all together form a big network that amounts to a second brain, one that is larger than the central brain if all the arms' neurons are included.

The biologist and roboticist Frank Grasso wrote a paper called "The Octopus with Two Brains," where he discussed this idea in a cautious way. Sidney Carls-Diamante, in a philosophical discussion, looked at what this might mean for the animal's experience. Perhaps an octopus has two different conscious streams, one for each brain.

Carls-Diamante discusses this alongside the 1+8 option, as two possibilities. The most thorough exploration of the 1+1 possibility I know is in a science fiction novel, *Children of Ruin*, by Adrian Tchaikovsky. The book imagines octopuses that have evolved into more technically intelligent animals (with the aid of human intervention). When describing each octopus, Tchaikovsky treats it in three parts, the *Crown*, *Reach*, and *Guise*. The Crown is the central brain. The Reach is the neural network in and across the arms. The Guise—not really an agent, just a part of the animal—is the color-changing skin. He describes octopus behavior and experience as an interaction, almost a dialogue, between the Crown and Reach, who have different skills and styles. It's hard to tell if the Reach is supposed to be an experiencing subject in the book, but it is something close to it.

Other possibilities arise in addition to these three—perhaps the unusual organization of the octopus takes it out of the realm of experience altogether. I will set that possibility aside, as the evidence for *some* sort of octopus experience is so good. What I am calling the "1" in 1+8 might also perhaps be two, the left and right sides. The view I will cautiously defend here, though, is a mix or combination of the first two options, the 1 and the 1+8. There is a way of putting them together, without paradox.

The idea I want to explore is a switching, back and forth, between a more unified and less unified situation. It's probably going too far to say that the animal switches between being one experiencing subject and being nine, but a less extreme version of this view might be right. This view was suggested to me—not with the details below but in outline—by Jordan Taylor, who was then a student, at a talk I gave at the University of Pennsylvania. I was comparing the 1 and the 1+8 views, treating this as a choice, and Taylor asked: Why couldn't the animal switch back and forth? I will explore this option by way of an excursion into another case, more famous and dramatic: the split-brain phenomenon in humans.

I introduced the basics of these cases above. We have patients whose bridge between the two brain hemispheres has been cut, to treat epilepsy. They often behave normally, but in some experimental circumstances they seem to have two minds in one body. This happens when different information is sent to each visual field, the left and right, as these are connected to different halves of the brain. (I will say "halves of the brain" for brevity, even though it is only the upper part of the brain that has been split—a point that will become important later on.) The connections have the crossing-over I described earlier: left visual field to right brain, and vice versa. In experiments, when different halves of the brain are shown different things and the person is asked to say what they are, all sorts of strange responses are seen. Speech is usually controlled by the left side, so the person will only talk about whatever is in their right visual field. But the right side of the brain, by controlling the left hand, can sometimes give its own answers, perhaps by pointing or doing a drawing.

The puzzle these cases raise is not just whether it makes sense to have two minds in one body, but the fact that normal behavior is the result, at least much of the time. Split-brain patients outside of experimental settings do not usually seem "split." They seem to go about their lives as ordinary whole people (not always—there are exceptions). What must be explained is the combination of apparent unity much of the time with apparent splitness in special contexts.

The debate about these cases has raised four main options (with many variations on each). First, it might be that there is only one conscious agent, one mind, present. This might be in the left brain. The problem with that view is the fact that the right brain can be quite smart in some experimental settings. The second option is that both halves of the brain are conscious—there are two minds. This view faces the problem that the patients behave apparently normally in most contexts. In this view, unified behavior arises from subtle coordination between the two agents.

A third possibility is that there is switching back and forth. This might be switching between a conscious left and a conscious right, but the option I will be looking at is a switch between having one mind and having two. Perhaps the special circumstances of experiments create dual minds, but the rest of the time there is one. I will call this *fast switching*. Lastly, there is the *partial unity* option. Perhaps there is no *number* of minds present in these cases. There is mind present in the bodies, but we are asking the wrong question when we ask whether there are two or only one; experiences are not always organized so neatly.

I think a version of the fast switching view might be right. I am far from certain about this; split brains raise so many puzzles, and individual cases differ one from another. But working through the fast-switching solution will also help with the puzzles of octopus experience.

To make a case for fast switching, I will first defend the view that in some situations, there really are two minds in one body. Here I draw on the philosopher Elizabeth Schechter's work on the topic. The best argument for two minds is based on cases of improvised communication between the two halves of the brain, communication that takes place out in the public world. For example, in a number of documented cases, the right side of a patient's brain, which cannot speak, has tried to convey messages to the left side by writing with a finger of the left hand onto the back of the right hand. If the right hemisphere knows the answer to something and the left does not, the right may try to get a message across in this way. This was sometimes an attempt to subvert the point of an experiment, as the aim was to show images to the right side and see how much the left could say about them. The experimenter would see the finger writing and say, "Don't write!"

This is a smart behavior, nothing like a reflex. It is good evidence for a mind in the right hemisphere, trying to get information to another. Sometimes there are two minds. And perhaps, as Schechter thinks, this is permanent. But that's not the only option.

Another possibility, again, is that much of the time, both halves of the brain work together to give rise to a single subject of experience, and sometimes this subject splits into two. It might seem dualistic to think that a whole mind can leap into existence, and go out again, as a consequence of relatively minor physical changes such as being put into a psychological experiment. But if a mind is a pattern of activity, then why can't it come and go very quickly, transform in shape, fall into one dynamic and then another? A harder question for this view is how the pattern of activity that makes up the *single* mind, the one present most of the time, can exist at all, given that the two halves of the upper brain are physically separated.

Early discussions of fast switching did not grapple with that last question very thoroughly. Susan Hurley, whose ideas played a role in Chapter 5 of this book, then offered a view that would make sense of the unity of a mind in a brain lacking its usual connections between the two sides. Hurley was not defending a fast-switching view, but trying to make sense of how a single mind could exist in split-brain cases. She said that some of the physical connections that unify the brain of a conscious subject might not be pathways that run within the skull, neuron-to-neuron, but might be pathways that extend out into the world and then back again, in loops. Tight and rapid feedback between the person's actions and what they sense might be part of what achieves this. The physical basis for a unified consciousness can exist, in Hurley's evocative words, as a "dynamic singularity" in a "field of causal flows" that extends some way outside the body itself. Another philosopher, Adrian Downey, recently showed that Hurley's view fits naturally with fast switching between one mind and two in split-brain cases.

Suppose we accept that something like this is possible in principle. Should we take it seriously? To show that we should, I will introduce another medical procedure.

The Wada test, named after its inventor, the Japanese-Canadian doctor Juhn Atsushi Wada, can be done on anyone (though it is invasive and not at all routine). The aim is to see which side of a person's brain controls language. The way it works is by putting each side to sleep with an anesthetic, one at a time, and leaving the other side awake. At each stage the patient is tested on their use of language, to see what they can do with one half of their brain asleep. In most cases, when the left side is asleep the patient can't speak. But the patient is still conscious, or seems so, at both stages:

During the test he would show me something, ask me to identify it and whether he had showed it to me before. . . . For the right side of the brain I didn't notice anything different. For the left side—wow! When he showed me an object I looked at it and had that feeling you get when you can't think of a word, like it's on the tip of your tongue. Only that was true for all words—it was amazing! I had no words.

Each half of the brain can apparently be conscious on its own, with the other half asleep. (Dolphins sleep one hemisphere at a time; they give themselves a natural Wada procedure.) I learned about the Wada test and its possible implications from the philosopher James Blackmon. He thinks the test shows something very surprising. As Blackmon sees it, if one half of the brain can be conscious when the other half is asleep, it seems that both halves must have already been conscious, in their own right, before the test. Each half evidently has what it needs to be conscious, and surely a half can't be changed very much by putting something else, something separate from it, to sleep. The half that is not asleep is just going along as it was. So Blackmon thinks each half must have been conscious all along. The consciousness of a normal fully awake human is a sort of blend or meld of the two consciousnesses of the halves, or perhaps a blend of many even smaller consciousnesses.

However, another possibility in the Wada situation is a kind of fast switching. It is true that putting the left side (for example) to sleep does not greatly alter the right side of the brain itself, but it alters the overall pattern of activity in the brain-plus-body. I think that when a Wada procedure is done, the remaining half becomes the *right sort of thing* to support conscious experience. Before the procedure, only the *whole* brain will support the right kind of pattern of activity. During a Wada test, the left side loses connection to a functioning right side, and the physical basis for mind "contracts" as a result. A pattern of activity that formerly extended across the whole brain is now confined to one half. That half is now conscious, but it was not a conscious system in itself beforehand.

At this point I should also acknowledge a complication. Following Blackmon's description, I am writing as if entire halves of the brain are put to sleep at once, but there is a "lower" part of the brain that is not split in split-brain experiments or put to sleep, by halves, in a Wada test. Similarly, the habit of talking about split-*brain* cases in humans began at a time when the upper part of the brain, the cortex, was thought to be responsible for just about everything in thought and experience. Now the role of the lower part of the brain—which is not split in these procedures—is taken more seriously. Ordinary conscious experience in humans might arise from the operation of both these parts of the brain working together. When the left or right half of the upper part, the cortex, is put to sleep in a Wada test, what remains active and associated with conscious experience at each stage is the other half of the cortex plus the entire lower part of the brain. This would still show what I am calling "fast switching," but the switch would be between two units that have some parts in common.

If this is right, then what the Wada test shows is that the physical basis for experience is a pattern of activity that can come and go quickly. Once we accept this kind of switching, it becomes more reasonable to say that in split-brain cases there might be a rapid switch between one and two conscious selves, and this can happen as a result of an experiment that separates the flows of information to each side.

This is not all that is going on in split-brain cases; I think there is another piece to the puzzle as well. When there are two minds present, the two are not wholly separated. There also seems to be some degree of partial unity.

"Partial unity," again, is the idea that in split-brain cases there is not a neat count of minds—either one or two. In some respects there is a separation, with experiences and memories partitioned in two, but other thoughts or experiences can belong on both sides. Many of the doctors and researchers who work with split-brain patients think this situation is often present, because they think that moods and emotions can be shared across the two sides, presumably through the lower, unsplit part of the brain. A single mood—being stressed, for example—can be part of mind 1 and also part of mind 2, and that means there are not really two distinct minds at all. The idea is not that both minds happen to be stressed in a similar way, but that a single stressed-ness is part of both. Researchers think this because when a stimulus is shown only to one side of the brain, the other won't know what it is, but might have its mood affected, even though that side can't make much sense of what has happened to its mood.

That completes an overall view of split-brain humans. There is fast switching between one mind and two; the one mind, when present, is unified in part by causal pathways that extend outside the body; and when the one mind splits there is partial unity with respect to things like moods and emotions, which means the "two" minds are not fully distinct. I have embraced both the more contentious options, fast switching and partial unity, and put them together. Neither is as difficult to imagine as people have often supposed.

Split-brain cases are full of complexities, and this interpretation might be wrong. In at least some cases, perhaps there is a permanent two-mind situation rather than switching (though I think the "two" will include some of that mixed-up situation with partial unity and shared moods). We will also revisit these questions in a later chapter, when another piece of the puzzle is on the table. But I think that working through this interpretation of the split-brain puzzles helps us think about experience, subjectivity, and brains in general. For example, let's return now to the octopuses. We are not talking anymore about the results of surgery or special circumstances, but about ordinary octopus life. Perhaps there is some of the same sort of combination of features here, in a milder form—a form without the dramatic coming-into-being of conscious selves. Here is what I have in mind.

Sometimes an octopus is a single unified agent. As it throws debris, jets around, and the like, the animal is unified and its experience reflects this. But at other times, the arms are allowed to wander and explore, and perhaps the central octopus does not "own" these locally guided motions.

When they roam, the arms are like very simple agents. They sense and they respond in action. Do they have experiences of their own? I think that is probably saying too much. They probably have enough complexity in there for experience—enough neurons. An entire bee has only a million neurons, and an octopus arm has tens of millions. But most likely, arms never become *selves* of the right kind, given their connections to the networks of neurons in the other arms and to the center. There is, after all, a 1+1 view on the table as well as a 1+8 view; this reflects the fact that the neural networks in each arm are not self-contained, even when (as it appears) the arm is allowed to wander and do its own thing. Perhaps neither the entire multi-arm network nor the individual arms have the right kind of integrity to be genuine subjects. But the arms do have *some*, a hint, of what I have been talking about as the basis of subjectivity. If there were to be glimmers of experience in individual arms, or some of them, the situation would probably also be one of partial unity. Stress, energy level, arousal, and so on might exist in a unified way across the whole animal, even when the arms are sensing and responding on their own. I should note also that although I have been writing as if all eight arms are equivalent, the first and second pairs of arms are generally more active than the others.

In any case, the arms lose autonomy when the octopus "pulls itself together" and imposes central control. Then these features of the arms, whatever exactly those features are, become submerged. The big shifts in subjectivity we have been looking at in the human cases (Wada, split brains) come about because something is done to the person. In the octopus case, if there are shifts that unify and disunify the animal in these ways, they arise differently. The octopus pulls itself together, perhaps as a response to a situation that requires focused action, by exerting control over what was being allowed to wander.

If this is right, how is this pulling together different from what happens in ordinary human cases, where a person suddenly pays attention to their own breathing, chewing, or walking after previously paying it no mind? How is it different from playing a tune on a musical instrument while paying no attention, and then suddenly bringing it into focus? I do think it is different. In the octopus case, when the center is not exerting control, the arms engage in activities that are agent-like, locally controlled by what they sense. The arm's own motions will also have consequences for what the arm senses at the next step. The human cases are more like the use, and occasional overriding, of a central autopilot.

Compare a situation where your hand is wandering, outside of attention, and then it touches something of interest. Information goes to the central brain, and unless this is a reflex or other special case, the hand won't do anything in response unless the brain directs it. In the octopus, we are not sure what exactly is going on, but the picture on the table is one where if an arm touches something, information does go to the central brain, but the arm also produces its own locally controlled response. The animal as a whole knows it was touched, and can see and perhaps feel what is done. But in these situations the arm itself is determining its response. So when the octopus exerts attention and control, this achieves more than it does in our case. In the octopus, it is pulling together partly independent processes in different parts of the body, in a situation where the parts will head toward a kind of agency of their own if left to their own devices. In this way, ordinary octopus life might include a fainter version of human split-brain dramas; it might include switching between whole-animal centeredness and a glimmer of autonomy in the arms, or some of them. That glimmer is submerged or lost when the octopus uses attention to direct a coordinated action. Earlier I worried that integration looks important to subjectivity and hence experience, but octopuses are not very integrated. Now we see them *becoming* more integrated, and less so, in real time.

The octopus puts pressure on so many parts of the picture I am developing. Maybe their 1+8 organization will one day force us to rethink the whole idea of subjectivity and its relation to integration in animals. Even with much unresolved, this discussion of split brains, the Wada test, and the octopus affects our picture of how the mental can exist in the physical in a number of ways. First, fast switching is not so extravagant an idea. If a mind is a pattern of activity, it might come and go, transform, or expand and contract very quickly. It is easy to express a kind of official acknowledgment of this point, but in the Wada test we see some of its consequences. The idea of partial unity has also been seen as highly dubious, but this, too, can be part of our framework for thinking about animals. Various animals have sensory streams that are somewhat separate from each other, but other states like moods, satiation, and stress might be shared across the whole.

If we set aside the admittedly irresistible question of tiny additional selves in the octopus, and assume the animal always retains a single center of experience, the consequences of the octopus body for this experience may still be quite exotic. An octopus can spend long periods sitting quietly like a cat, perhaps close to sleep. Sometimes, in contrast, they are intensely active: jetting, throwing, building. At yet other times you see something in between, and these are especially intriguing moments for questions about the experiencing self. An octopus might be wandering over easy terrain, and its arms will then move in a way that generally looks centrally directed, but combined with wandering detours and ambulatory ornamentations. An octopus may sit in place while two or three arms explore in different directions around it simultaneously. Though an octopus guides many of its actions with vision, it is also extremely sensitive in other ways. The suckers on each arm are packed with chemical sensors. When an octopus's arm touches your finger, the arm is *tasting* it; an octopus perceives a huge difference between a human touch with a thin glove and with bare skin. The skin of the entire animal also appears to be sensitive to light. It is going too far to say that an octopus can *see* with its skin—its skin cannot form and process an image—though not only the intensity of light, but also changes, shadows, and perhaps hues, might be detected with the entire body.

Putting all this together, we reach an experiential situation far from our own. Assuming that sensory information from skin and suckers does get to the central brain as well as to local neural networks, the octopus becomes an animal with both a very expansive sensory surface and, from the central brain's point of view, a rather unpredictable one. As the arms wander, they will change the shape of the body and also encounter objects, surfaces, and chemicals that produce sensory events. This can happen in several arms at once. The octopus does *occupy a perspective*, but a protean and perhaps sometimes chaotic one. When I try to imagine this I find myself in a rather hallucinogenic place, and that is everyday life for an octopus.

Down Among the Stars

One day in the middle of winter I swam down an anchor line to Octopolis with Matt Lawrence, who had discovered the site ten years earlier.

It was a bright day up on the surface, but as we went down the light faded and grew dim. Soon we were just above the flat plain that surrounds the octopus site. When we reached Octopolis, we saw four sharks right around the site, all with their noses pointed in as if forming a perimeter. These were the same kind of shark described in "Octopus and Shark," but all were much larger. One was at least eight feet long, perhaps longer. There were a few octopuses around, but they were well and truly keeping their heads down, so Matt and I went wandering over the seafloor.

Soft gray sand covers the floor of that part of the bay, with seaweed of different kinds in olive and dark green. As we headed out I soon noticed a lot of starfish, many more than usual. Looking closer, they were feather stars. Feather stars are often not called "starfish," as that name is reserved for their thicker-armed relatives. But both are *echinoderms*, with the same star-shaped design. Feather stars, as the name suggests, have fine arms like slender plumes.

Echinoderms contain two main groups. *Crinoids*, which include feather stars, make up one, and most of the familiar starfish and urchins are in the other. Crinoids feature both the least and the most mobile echinoderms. Some are attached permanently to the seafloor with a stalk—sea lilies. Others can actively swim.

These ones near the octopus site were small, with arms just a few inches long. Most were white, or silver-white in some cases, along with a few that were dark purple. They weren't absolutely everywhere, but you could not go for more than a few yards without one appearing in the dim light. We were swimming through a field of stars.

In our journey through animal histories, this marks the point where we make another significant step, crossing over from one branch of the tree of life to another. Among animals, there is a big branch of bilaterally symmetrical species that we moved onto when we left the corals, with their disc-like form. That branch made up of bilaterian animals has two main sub-branches, in turn. Arthropods like crabs and molluscs like octopuses and slugs are on one side—these are in the *protostomes*. The other side is the *deuterostomes* (pronounced "dew-ter-o-stomes"). This is our side, the side of vertebrates. With the exception of unwelcome visits by sharks, and some fish and ascidian cameos, we've not ventured onto the deuterostome side so far in this book. From now on, that is mostly where we'll be. The animal that greets us on arrival, though, is a surprising one: the starfish.

Echinoderms have been around at least since the Cambrian. Reconstructions of their first forms have them as flattened crawlers, like the actual and imagined bodies of various other early bilaterians. They evolved into a range of lopsided and then spiraled forms, followed by their discovery of that star design, a design that seems a gesture back to the soft corals of Chapter 3.

A few starfish can crawl quite quickly, and I have been startled to see large sea cucumbers, their bodies facing upward, stuffing their mouths with food in a great frenzy of consumption. But many echinoderms have settled into a slower pace of life and a flower-like body.

We swam through a field of these animals. Not everything down there was sidereal. As we reached a spot that is often an outpost of the octopuses, a big seal swam in. It came in very fast, close to us, pausing and peering and zooming away. It showed astonishing agility, curling and surging through space.

The seal came in so close that I thought it was going to head-butt me. Then it turned again and was up and gone, toward the surface and the light, leaving us down there among the stars.

7

KINGFISH

Power

Toward the end of a dive on the Pipeline site at Nelson Bay, I was making my way to shore and paused in the last deeper place before the shallows began. Then my head was turning before I realized it, and I was inside a thrumming cloud of missiles. They looked about a yard long, a few dozen close together, with silvered metallic sides and slender sickle-like tails. They were yellowtail kingfish (also known as king amberjack). Their tails seemed too delicate to generate the power that was coming from them. Sheer vertebrate motor power.

I was in the middle of the group and they turned in one brief arc, a crescent like their tails, and were gone. I felt I could almost *hear* the electricity in their muscles—the surges of ions, the effortful contraction of cells. This sense of a sound was an illusion, and the sea was still quiet; instead, I was feeling the shuddering displacement of water around me.

Various elements of the animal kingdom are celebrated for speed or power, some in animals that we have come across

already: the firing of tiny harpoons from stinging cells in jellyfish, the spring-loaded hammers of mantis shrimp. It's a different thing, however, to generate fast motion on the scale of a
big fish, to propel a body of that size through thick water. In
those kingfish we are seeing the triumph of muscle, and of the
vertebrate design, with muscle laid on bones. Fast fish are one
pinnacle of animal motive power.

History of Fishes

Fish have a special role in this evolutionary story as they are in
our group of animals. Or, more accurately, we are in theirs.
Mammals are an offshoot of the fish part of the tree of life. We
are descended from fish in a way that does not apply to animals
encountered in earlier chapters. In a book by the biologist Neil
Shubin, we are asked to give thought to our "inner fish," as a fish
lies in our ancestry, and the anatomy of a fish underlies much of
our own. In this sense, we do not have an inner shrimp or an inner octopus.

We are not descended from a fish that looks like a kingfish or
trout. Those are later arrivals, appearing after our line branched
off. Instead, our closer relatives are more awkward-looking,
stubby fishes, the lobe-finned fishes, represented today by the
deep-sea coelacanth.

Fish began very much as minor players. They appeared in
the Cambrian as inconspicuous slivers, an inch or two long. At
the end of the previous chapter we crossed over to the deuterostome branch of animals, a branch that contains echinoderms, us,
and a few others. These animals began, like many others, as one
wandering path from early worm-like bilaterians. At some stage,

an animal of this kind evolved a more mobile, slender form. It was unable to do much; it did not even have teeth. It was a nondescript scrap in the sea (*Pikaea*, pictured as animal B on page 82). But this new animal had some distinctive features, including a missile shape, a nerve cord down the back organizing muscle, and the beginnings of hard parts, not on the outside, as in arthropods, but on the inside.

In the Cambrian, when arthropods evolved predation, these fish were among the likely prey. Probably for this reason, fish were quick to evolve good eyes. These eyes had a "camera" design, with a single lens and retina, rather than a cluster of them as in most arthropods. Early innovations in vertebrate animals included a body made for mobility and a camera eye.

From the Ordovician onward, fish began to get larger. They soon included animals with bony plates, and up to a meter long, evading predation now with size and armor. Some were fearsome-looking, but they were not fierce in fact. Despite an imposing body, they had no way to bite, and probably scooped, sucked, and filtered. The decisive invention that followed, some time over 420 million years ago, was the jaw.

In a classic example of what the French biologist François Jacob has described as evolution's "tinkering," jaws were fashioned from gill arches on the sides of some fishes' heads. This was tinkering with profound consequences, as the result was a unified, muscular bite. In an apt comparison (owed to Jane Sheldon), a jaw is like an opposable thumb for your face. With the combination of swimming, eyes, and jaws, fish took on a new role. By the late Devonian, around 360 million years ago, jawed forms had diversified and jawless fish were fading. The ones that hang on today live on the fringes—hagfish, lampreys.

Devonian jawed fish included sharks. As John Long says in

The Rise of Fishes, sharks settled early on their design and have merely fine-tuned it since then. In 2018, an amateur fossil hunter in Australia found some unusually large teeth. Immediately recognizable as shark teeth, they were still very sharp, 25 million years after their owner, a nine-meter-long whale eater, had died.

Swimming, eyes, and jaws create a powerful combination. These were married to another feature that seems to me a notable vertebrate inheritance. Their body plan and ways of moving make fish particularly *centralized* animals, and this is reflected in the vertebrate brain. Here, as elsewhere, surprising disunity can be hidden; the fish brain has two sides, with a good deal of separation. But in comparison to the other bodies we have encountered so far, this is a centered, centralized form. The brain guides whole-body actions—just about all a fish *has* are whole-body actions, with no arms, claws, or tentacles. Later, when vertebrates acquire manipulating limbs, they are controlled from the same kind of brain.

I once saw a kingfish hunting. I was watching some squid in the shallows. They wandered about, eyeing me as I slowly pursued them, occasionally signaling with skin patterns. Suddenly a large kingfish appeared, again with that sickle tail. It was among us in an instant, moving in a jerky, staccato manner. It had no limbs, none of those squiddic appendages, and its movement was a whole-body stab. The squid, scattering, emitted gouts of black ink that hung in the water like puffs of anti-aircraft fire.

Swim

I think of this as the blue-water chapter, the chapter that embraces blue water. Much of the scuba diving I have done while writing this book has been among tangled thickets of life, and has proceeded at the pace of crawling and clambering animals. I am slow in my gear, but crabs, corals, and even octopuses, most of the time, are not too fast either. Squid are an exception. But a lot of the path so far has followed lives of detailed engagement with small locales.

Compare this to fish, with their vast journeys under the surface of the planet. These are animals that take an ocean as their scene of action. Many take *more* than an ocean, being born in one, migrating to feed in another, and migrating back to breed, feeling magnetically the whole of Earth.

I began the chapter with kingfish. But in the domain of vertebrate power, the most extreme form I've seen is the tail of a whale shark. The whale shark is the biggest fish in the oceans today. *Rhincodon typus*—the only species in its genus, the only one in its family—is also the largest animal alive other than some whales. Individuals reach at least forty feet in length. Their bodies are dark gray, with white spots and bands of several paler grays. The body is propelled by a huge vertical tail.

The triangular top part of this tail, about six feet high, looks like a sail underwater, though unlike a sail it makes its own power. In front of the tail is an enormous body with heavy, ribbed lines running back to front. These look like welds on the body of a plane.

Whale sharks travel around the world's tropical seas. No one has seen them mate. At the time of writing, it is suspected that they all mate at a particular place, near the Galápagos Islands. If so, from there they move on to Australia, Mexico, the Philippines. They cruise near the surface and feed on plankton, as many whales do. As whale sharks often swim slowly at the surface, you can snorkel with them, as I did in Western Australia, near Exmouth.

You swim alongside and may also find yourself above them. The first time those body patterns appeared below me, it was like hovering over the surface of a planet. As they swim, a cloud of smaller fish often swims under, alongside, and within the open mouth of the shark. Those are not being swallowed, but just traveling along in that cathedral mouth. The constant busyness of these companions contrasts to the still self-possession of the shark.

Occasionally a shark dips down to dive. The head slowly tilts and the animal heads into darker water. They can dive over a mile deep. Unlike a whale, they have no timetable of return; they are not mammals needing to come up to breathe. The couple of times I've been in the water with whales, I have had a sense of them as mammals like me. Whale sharks, breathing water, are *of* the sea in a deeper sense.

Swimming on a clear day with them, twice I saw a whale shark tilt downward and head below. Soon he was perhaps thirty feet down, but we could follow easily from above. Passing over a reef in clear water, I saw afresh the grace of his motion, with the almost gentle swing of the tail, over and over, propelling the body forward.

This, like the kingfish, is a triumph of muscle, and of the vertebrate design with muscle laid over an internal frame. Sharks do not have bony skeletons, but use lighter and more flexible cartilage. They do make bone for a few parts; bone is available to them. The transition to bonier vertebrates may have come with the duplication of a bone-making gene, allowing bone to acquire a wider range of roles. Most of the familiar fish we encounter have skeletons made of bone.

Not far from where the whale sharks appear, I saw a cleaning station for gray reef sharks. This is a spot where small fish wait and large ones come in—sharks, in this case—"clients" to be cleaned. The small fish eat tiny parasites from the client's skin and inside its mouth. As well as removing parasites, they take a small extra nip, very often, from the client itself. Clients accept this, in moderation. The sharks at this site tended to give a small, quick shudder at the end of a clean, probably a response to the cleaner's extra nip.

The cleaning station was situated over some huge dome-like mounds of coral. The reef sharks came in quickly, moving by no apparent means. They seemed to transmit themselves through the water with almost imperceptible movements. Quite different from the whale shark, and the rhythmic swing of a self-propelled six-foot sail.

The Presence of Water

Fish, from early days, had good eyes. They had the old inherited chemical senses. Fish evolution also gave them—again early, from the days of jawless fish—a more distinctive sensory invention, the *lateral line* system.

This is a tactile sense, roughly speaking, a form of touch. The main elements are small collections of cells with protruding hairs, each collection inside a soft cup. These units, spread in many places over the body, are called *neuromasts*. They are sensitive to the movement of water and make contact with nerves beneath.

Some of these neuromasts are on the outside of the animal, while others lie in thin, fluid-filled canals just below the body surface. The canals, running along the fish in many directions, but mainly from front to back, are open via pores to the water outside. The neuromasts inside a canal can be sensitive to subtle differences in pressure and tiny vibrations. A water current moving over the body creates pressure differences as it interacts with different pores.

I said this system was a form of touch, but the motions and pressures of touch shade into sound. Hearing is the detection of pressure waves that move parts of the body. In us, this movement runs from the eardrum through to cochlear hairs in the inner ear. In fish, the lateral line system mixes touch with hearing. The evolutionary story goes something like this. Hairs protruding from cells are very old, and were probably present in our single-celled ancestors. Those hairs can be used both for propelling the body—an active role—and for sensing, registering touch and motion in a surrounding medium. In animals, the gravity-sensing organs of jellyfish we briefly bumped into in Chapter 3 work like this, with hairs detecting the motion of crystals as the body tilts. In vertebrates, it's possible that lateral line canals were precursors of ears like ours, but also possible that motion-detecting hair cells were put to work in our ears through a different pathway.

Fish themselves hear well. They have ears of several kinds, along with their lateral lines. Some also assist their hearing with swim bladders, gas-filled sacs whose main role is as a buoyancy device.

The lateral line system has been described as "touch at a distance." It detects movements, both close by and farther away, and as it encompasses so much of the fish, it must give rise to a strong bodily awareness. We land mammals have more separation, over gaps of air, from most environmental events. Water conducts vibration and movement very readily, and lateral line canals are open to the sea outside. The result is touch-like contact with much of a fish's environment. Think how it is when your ears fill with water as you start to swim. In an aural gurgle, the air in your ears is replaced by water and things get louder and closer. A nearby boat engine is felt as well as heard. The lateral line system is like that, but it extends over the whole body. Octopuses, as we saw, have some light sensitivity over much or all of their skin. It would be an exaggeration to say that their body is a huge eye; the system is almost certainly too simple for that, and cannot form an image. It is not as much of an exaggeration to say that a fish's body is a giant pressure-sensitive ear.

Lateral line sensing features rich interactions between sensing and action, of the kind I discussed in Chapters 4 and 5. As a fish moves, it affects the surrounding currents that it also feels. The movement of objects in water leaves long-lasting traces. Even a small fish leaves a wake that can be detected minutes after it has passed by. To make sense of all this, the fish must filter out the effects of its own motion from the effects of everything else. Some of this is done right away; the nerves linked to hair cells in the lateral line system include some that inhibit the sensory signal when the fish is creating its own disturbances in the water. As I also said in those earlier chapters, the effects of one's own actions on the senses do not only pose problems, but also present an opportunity to probe the environment, to learn more than you could by passively taking things in. Blind cave fish, as the name suggests, are entirely blind, and navigate with lateral line sensing. They actively generate motion in the water that will interact with nearby objects, and can use information of this kind to pass through a barrier of rods without touching them—steering by a kind of lateral-line sonar. These fish swim notably quickly when faced with unfamiliar landmarks, apparently to feed their lateral line systems more information.

In some fish, especially sharks, the lateral line system has been modified to enable another form of sensing; they can sense electric fields as well. Fish of various kinds use both passive and active electrosensing. The difference is whether you send out electrical impulses of your own, or just detect the ones that arise for other reasons. Jawless fish and sharks engage in passive electric sensing. Sharks can detect fish hidden in sand by feeling their natural electrical activity from above. The function of the peculiar head shape of hammerhead sharks is unclear, but one role may be to improve electrosensing. The shark researcher Aidan Martin described seeing hammerheads swimming close to the seafloor and swinging their heads from side to side in arcs as if wielding a metal detector: "On several occasions, I have seen 'mine-sweeping' Great Hammerheads double-back suddenly to scoop up one of several stingrays buried in the bottom silt."

Interestingly, this electrical sensing has been generally lost in the bony fish that radiated during more recent periods of fish evolution. A few of these fish later reinvented it, especially catfish, whose form of electrosensing seems to be so sensitive that it predicts earthquakes. Some bony fish also engage in active electrosensing, producing their own electric fields and detecting nearby objects by alterations in that field.

A few fish, especially rays, can generate considerable electric charge and use it antagonistically, rather than for sensing. Some years ago while diving I unwisely put my hand down on what looked like a small and bare patch of sand. I felt immediately a THUMP. I thought it was an impact—I thought something had nudged me pretty hard. I was turning to look, and then right away: BANG. Much harder, and definitely an electric shock. Sharp and precise, a discharge, and not a good experience. It was a numbfish or coffin ray, buried in the sand and sending me on my way.

Setting aside electrical elaborations of the lateral line system, its touch-like side has intriguing relations to another fish behavior: schooling.

Schooling in some fish is truly uncanny to watch—the sudden shifts in direction, where leader becomes follower and follower becomes leader, seem instantaneous. Seeing this in the sea immediately raises suspicions of hidden fields or a giant mind. The school seems to make an instant group decision, with hundreds of fish turning in the same direction at the same moment. Despite appearances, this all occurs through the very rapid perceptions, decisions, and actions of individual fish.

Once I learned about the long-distance touch of the lateral line system, schooling immediately seemed less mysterious; it seems to be exactly what that system would enable fish to do. What surprises me is that research looking directly at the role of the lateral line system in schooling is equivocal. Some writers argue that vision is responsible for nearly everything in schooling—a claim that restores my bafflement at what happens in the sea. A number of fish species can school, at least to some extent, after having their lateral line system disabled, while others do seem to rely on it. Perhaps the most integrated forms of schooling work differently from others. Not all schooling involves microsecond high-speed coordination.

I have spent such a lot of my time in the water peering at invertebrates, and it has been partly the process of writing this book that has brought me into closer contact with fish. Once during early work on this chapter, I was on a cold winter dive at a site near Sydney. I came across unusual numbers of a fish I'd seen there often, flutemouths. These are extremely, almost comically slender, often about two feet long. I followed four of them for a while.

Over a quarter of the body is head and snout, about half is the body proper, and the rest is a thread-like tail that extends behind. There are two almost invisibly small fins up near the head and another pair farther back. These are nearly transparent, and about the size of matchbooks on a two-foot long body. A pair of tail fins are even smaller, and then the silver thread stretches behind like a long antenna. Those tiny fins are used to gently push through the water. Flutemouths are creeping predators, edging toward prey imperceptibly, with those tiny fins both hard to see and, I assume, hardly disturbing the water. Then they strike.

They are quite elusive in the water; you see them and then they are gone, though not far. Sometimes I'd lose them and they would reappear like four lines drawn calligraphically in the sea—drawn suddenly at one combination of angles, then another. Reading about them afterward, I found that the antenna-like tail contains an extension of the lateral line. What looked like an antenna *is* an antenna, one tuned to faint vibrations, tremors, and vortices in the water. Just after I'd started trying to think my way in to having a body like a huge ear embedded in dense water, I come across a fish wearing its lateral lines on a thread like a radar tower.

Earlier in this book, I imagined the sensory worlds of octopus and shrimp. The shrimp fills neighboring space with a body of protruding hard parts. These include antennae—a half dozen canes or walking sticks, but much more sensitive, poke out in all directions. The octopus has soft arms with their own intense sensitivity; everything touched is tasted, and the response to what is tasted comes in part from the arm itself. Now we reach fish. Here the sensory paradigm is not one of probing, or gustatory exploration, but the feeling of motion, the perception of passage through water, accompanied by a pervasive long-distance touch that blurs into hearing. Motion is the original fish métier— motion with the freedom of flight through three-dimensional space, and with its effects on the listening body of the whole.

Other Fish

We have moved through several stages in the history of fish, beginning with inch-long slivers in the Cambrian, proceeding through larger armored forms, then to the transforming jaw. Somewhere in this sequence, some fish also became smart. They did so in a way that initially looks a little puzzling, given the themes of this book.

A couple of chapters ago, I noted that the only animal to pass a version of the mirror test for self-awareness that is neither bird nor mammal is a fish, a cleaner wrasse. Fish will count objects in experimental tasks. They use counting as a last resort, apparently, using other clues if they can, but the same is true of dolphins and people. Fish have learned to discriminate different styles of music—blues from classical—and could extrapolate from one blues artist to another; they were not learning the quirks of one performer. That is a quite abstract feat of pattern recognition. A lot goes on inside a fish.

What is initially surprising is the fact that there is all this complexity guiding a body with so little it can *do*. The story I have developed here about some aspects of the history of the mind is one that places considerable emphasis on the evolution of action. In fish there is movement, and the all-important jaw, but they are behaviorally limited animals—not very manipulative, for example—in comparison to various animals of earlier chapters, especially arthropods and cephalopods.

Why then did fish (or some of them) become so smart? The question must first be asked the right way. The wrong way to ask it is: "Why do fish need to be smart?" This is not a question of need, but of relative advantage. If you are a fish, can you do a bit better than others in your population if you are a bit smarter, especially given the costs of building and running a larger brain? If you can indeed do better, what gives rise to this advantage?

Much of the answer seems to be that fish, even more than seems immediately apparent, are gregarious animals. They are continually interacting with others. Social interaction creates a complex environment for an animal, and is very often a driver of the evolution of intelligence. This principle was originally developed for primates, where especially large brains are found in the more social species, but it has a broader application, and fish look like a likely case.

Most known fish hang around with other fish, at least for part of their life. They prefer their own species and often prefer their kin. Fish in many cases can recognize individuals, and in some species fish prefer to form shoals with others they are familiar with. On my last scuba dive before sending in the manuscript of this book, I saw, under a small ledge, fish of four different species resting together so closely that their bodies in several cases were touching. There was a large catfish, a moray eel, a spotted yellow-green cod, and a spiky, craggy scorpionfish. The situation was not otherwise crowded and plenty of other spots seemed available, but that is where they wanted to be. My arrival was a different matter; pretty quickly the cod got up, looking irritated, and went out past me. Would many other wild animals do something like this, huddling with individuals of three other species? Jean-Paul Sartre (who, incidentally, had an intense and drug-amplified fear of crabs, octopuses, and some other sea creatures) is famous for writing, "Hell is other people." In the case of fish, it appears that heaven is other fish.

The cleverness of fish shows up especially in their dealing with others. An example is imitation. Imitation itself is fairly rare in animals, and it is rarer to imitate selectively—to imitate individuals who are doing well, rather than indiscriminately. Some stickleback fish can do this when working out where to look for food. Perhaps the most remarkable case of fish imitation, though, is seen in archerfish. These fish squirt water at bugs who are out of the water, knocking them in to be eaten. Usually the bugs are stationary, but archerfish can learn to hit moving targets. Remarkably, if one archerfish in a group learns to do this through practice while other fish watch, the watchers pick up the skill too, and hit moving targets, when given the chance, nearly as reliably as those who had practiced it. As the researchers say, this is especially impressive for the way that observing fish seem able to "change their viewpoint"—transforming the angles they see in a shooter into angles they produce themselves, when it is their turn. This is one of the most surprising animal experiments that I know.

I have mentioned cleaner fish a few times—small fish who eat parasites from the exterior of other fish who offer themselves for this service. Cleaners often take a bite of the "client" itself at the end of the cleaning, and I described what looked like a situation of this kind at a shark cleaning station earlier in this chapter. If a cleaner takes too big a bite, that is cheating on the arrangement, but small bites are OK. At cleaning stations, there is often a queue of fish, or at least others around, waiting their turn to be cleaned. One kind of cleaner fish tends to cheat less if there are onlookers, and cheat more if there are not. Individual fish seem to be tending to their reputation. And onlooking clients do indeed avoid cleaners who are seen to cheat.

This social tendency in fish is not limited to their relations with other fish. Fish have a broad interest in collaboration, and often work together with animals that can do more, or at least can do different things than they can. Quite a lot of the complex things that fish get up to seem aimed specifically at overcoming their bodily deficiencies in the realm of action.

On many seafloors you may find shrimp-goby collaborations. A goby is a small tube-shaped fish, and individual gobies often share a den, at very close quarters, with a single shrimp. The den is dug into the sand by the shrimp, with its Swiss Army knife appendages, and the fish is employed as lookout. The combination of arthropod and vertebrate features here is striking: the camera

TOP: Two large nudibranchs, *Armina major*, in front of some of the soft corals that open Chapter 3. Nelson Bay, Australia.

BOTTOM: Tube sponges form a choir at Lembeh Strait, Indonesia. These are not glass sponges, but rather demosponges, a more common group.

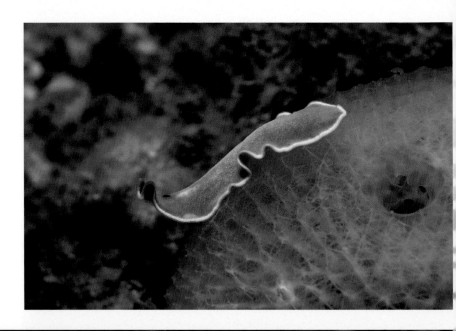

TOP: A polyclad flatworm, *Cycloporus venetus*, moving from right to left on a sponge. A tiny arthropod, almost invisible, walks behind it.

BOTTOM: *Phyllodesmium poindimiei*, an aeolid nudibranch, at Nelson Bay.

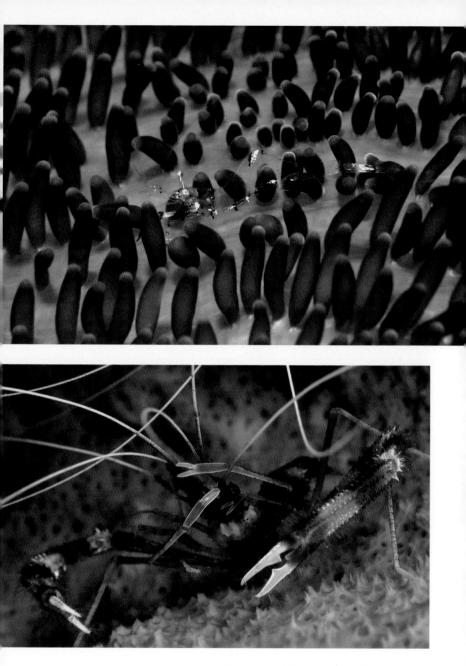

TOP: A transparent anemone shrimp, *Ancylomenes holthuisi*, on top of an anemone at Lembeh Strait.

BOTTOM: A banded shrimp, *Stenopus hispidus*, also at Lembeh, showing its array of appendages.

The darker octopus in the photographs above is described at the end of the section titled "Octopus Watching." The species is *Octopus tetricus*, here at Nelson Bay.

TOP: A giant cuttlefish, *Sepia apama*, at Cabbage Tree Bay, Australia.

BOTTOM: This is the octopus whose rampage opens Chapter 6, here seen addressing a soft coral with a predatory embrace.

TOP: A catshark and a hawkfish resting on a large sponge at Fly Point, Nelson Bay. The shark is the same species, *Brachaelurus waddi*, that dozes in the section titled "Maestro."

BOTTOM: A White's seahorse, *Hippocampus whitei*, also at Nelson Bay.

THIS PAGE AND THE NEXT: Whale sharks, *Rhincodon typus*, at Ningaloo Reef, Australia.

eye of the fish, a great sensory invention, makes it worthwhile for the shrimp to share its home despite the fish's uselessness as a builder.

An elaborate form of cooperation with a similar theme has been found in some large fish, groupers, in the Red Sea and on Australia's Great Barrier Reef. These fish hunt cooperatively with moray eels, and coordinate their actions with signals between the species. A grouper will often find prey hiding in some crevice in a reef. Being a large fish, there is not much it can do to get to that prey. It has no octopus arms or crab claws. But using headshakes and other gestures, it will signal to a moray eel that can go in and extract, or drive out, the prey. Alexander Vail, who has studied this behavior in detail, has seen fish wait above prey for up to twenty-five minutes before finding a partner to signal to. Vail also found that these fish had very good memories for him. He had fed a grouper a few times, and then was away from that spot for three weeks. When he returned, the fish came in unusually close and sat there waiting. Vail thinks it recognized and remembered him as a source of food. Though the usual hunting partners of the groupers are eels, octopuses have several times been seen responding to a grouper's signals and delving into the reef after prey.

A general story about the evolution of braininess in fish might run as follows. Fish are gregarious, as this helps them in various ways, but especially in dealing with predators. A complex social environment makes it worthwhile to develop recognition, memory, and strategic skills. These are then available for use in other contexts, and are also seen when fish find themselves in unnatural circumstances learning things of little biological use, such as the differences between musical styles. Fish also show an unusual amount of cooperation across species, even across the considerable fish-shrimp evolutionary divide. On the fish side, this arises because it gives more-cooperative fish an advantage over other fish in the same population. (The same is true, on the other side, for shrimp.)

All this makes some sense of fish braininess. I want to put one additional idea on the table. This idea is not about the cause of their braininess, but about the form of braininess they ended up with, and its consequences downstream. The fish body is a single whole, not a mass of outlying parts like an octopus. It is built for motion, and for other actions that involve coordination of the whole body. The nervous system is centered in the head between those camera eyes.

Often in evolution, an animal will be set up in a particular way for one set of circumstances, and then may find itself in a very different setting. When this happens, you will find yourself with a way of being, a way of relating to things, that is inherited from the earlier context. This might bring advantages or problems. Things will present themselves in a particular way and particular roads will be open—or relatively open—while others are not. Much was to come later for the centralized vertebrate brain, pioneered by fish.

Rhythms and Fields

Hans Berger, in his laboratory in Germany in the early 1920s, believed in telepathy, or ESP, and was determined to find out how it works.

He believed in telepathy because of an incident that had occurred while he was serving in the army some years before, in

1892. He was thrown from his horse and barely escaped being run over by the heavy wheel of an artillery gun carriage. That evening, he received a telegram from his father, as Berger's sister had reported a strong feeling that something terrible had happened to him. Berger was convinced that this showed some sort of telepathic connection—his terror at that moment with the approaching gun carriage had somehow reached his sister far away. Uncovering these hidden relationships between mind and matter became his passion in research. He did much of this work alone, often in secrecy, though he had a respectable position in a clinic at Jena and also carried on more conventional work. A younger colleague, Raphael Ginzberg, who became a subject of some of Berger's experiments, described him later as a person of obsessive routine: "His days resembled one another like two drops of water. Year after year he delivered the same lectures. He was the personification of static." Not entirely "static," though; a better term for someone who always gives the same lectures is "cyclical."

Berger sought to understand the mind-brain connection by looking at the movement of energy and its conversions into different forms. He tried various avenues, and eventually, knowing that another researcher, Richard Caton, had measured electricity in the brain, started using a galvanometer to measure electrical activity around the brain's surface, sometimes directly on the brains of people with injuries, and sometimes through the skull and scalp. Eventually he began to see something striking: regular waves of electrical activity, produced by the brain but detectable a little distance from it. Two kinds of waves became conspicuous, a slower and deeper kind seen when a subject had his or her eyes closed, and a quicker and shallower kind when the eyes were open. He called the first an "alpha" wave and the second, "beta." Berger had begun to take EEG recordings. He was not the first to do so—as often happens, history has uncovered near-forgotten precursors. But Berger was the first to do it in a human, and he gave his "brain mirror" its now-standard name—in German, the *Elektrenkephalogramm*.

The truth behind what Berger saw, or the truth as we currently understand it, is not telepathy, but is still rather strange. Puzzles surround it even now. The basics of what was going on are like this.

Electric charge and its role in living organisms were explored in Chapter 2. Ions are charged atoms or small molecules, positive or negative. Their to-and-fro is part of the endless traffic across the membranes of cells, and those movements are central to the taming of charge by living organisms. Nerve cells undergo action potentials, sudden chain-reaction spasms, by opening and closing their ion channels. The transmission of influence from one neuron to another is (usually) achieved by sprays of chemicals sent across the cleft between those neurons, to encourage or inhibit an action potential in the next cell. In a standard summary, brain activity is a big interconnected network of these events.

But electric charge has other roles as well. Across the membranes of neurons, ions go back and forth in slower, sometimes rhythmic, currents. These are in addition to the sudden flows that are part of an action potential, or "spike," described above. The slower movements do have an effect on action potentials, as they affect the cell's overall charge at each moment. Some of these slower flows are part of the action of one neuron on another, and others are "intrinsic" or self-generated, akin in some ways to an electrical breathing, a background activity on which is laid the drama of the spikes.

Suppose we zoom out from this activity and try to detect it from some distance away. Whenever a lot of electrical events are going on, their consequences tend to extend some distance in space. This is a result of a duality in the role of charge as a natural force.

Electrical activity has both a local side—seen in currents and chemical reactions, the side we've been looking at so far—and another side as well. This second side involves *fields*, extending invisibly though space. A field is a spatially extended pattern of some kind that exerts influence on things within the area it covers. An electric field is one kind of field. When something has electric charges at work within it, as a brain does, fields are generated around it, stronger at close quarters and weaker farther out. The neurons in a brain engage in tiny and continual electrical activities, as I described above. Suppose we could listen in to this activity in combination, all added together, and we listened from around the surface of the skull. We might expect to find a sort of random crackle or buzz. But we don't; instead we encounter coherent rhythms, and not just one rhythm, but several kinds, as detected by Hans Berger, searching for telepathy.

I said a moment ago that brain cells are doing a kind of electrical breathing. What we learn now is that some of them are breathing together—not all of them, or even most, but some of them, enough for the rhythm to be discerned if you attend to the whole and filter out the random activity. The rhythms are complicated, with waves upon waves, but they are there. The fact that we can listen in is due to the fact that those local electrical events generate a field, and the field changes with the to-and-fro motion of ions over membranes. (I find this auditory metaphor—listening in—irresistible, but in fact the rhythms are usually displayed visually on a screen or chart.) The patterns in an EEG are mostly due to those slower changes, rather than action potentials, though each affects the other. Berger described two wave patterns: alpha and beta. Since then, several others have been added, including the faster gamma waves and several much slower ones, seen in sleep.

These rhythmic patterns seem something like a signal. Over and above the local signals between one cell and another, an apparent transmission is generated by the whole. A transmission to whom? For what purpose?

The first possibility is that all of this might be a meaningless byproduct—an interesting accident, the incidentally musical hum of a machine. For a while, neuroscientists tended to assume something like this. But a lot has to happen to get a rhythm to exist, and these rhythms are found in a wide range of animals, including some very far from us. It's not possible to give an invertebrate animal an EEG of the kind given to a human, with a collection of detectors spread over the person's skull in a mesh. But it is possible to insert an electrode into the brain and hear the activity of cells nearby—not the millions of cells contributing to an EEG, but hundreds or thousands of cells. (This is called an LFP recording, for *local field potential*.) When this is done with fruit flies, crayfish, octopuses, and many other animals, rhythms are found, often in varieties similar to those seen in humans. Some rhythms are associated with sleep, others with attention, and so on. Octopuses have some especially human-like patterns. Given that similar-looking rhythms are seen in such different brains, it's hard to believe that the whole phenomenon is a meaningless accident.

A second option is that one of the things seen here is biologically important, while another is not. The important thing is the synchronized activity of cells, the fact that they do things together "in time." The larger electric fields that result, extending through space, and the wave-like changes in the fields themselves, are mere byproducts of this cell-level activity, with no role of their own in the brain.

Some of the same kind of reasoning used above initially seems to cast doubt on this view, too. In order to get waves in an EEG, we need not only synchronization of the activity of cells in time, but a lining-up of those cells in space. If all the cells contributing to the EEG were jumbled in space, pointing in many different directions, then even if they were behaving rhythmically, we would not see waves in an EEG, because the effects of the cells on the field would cancel each other.

That suggests that the fields and their patterns might also exist "by design," with a job of their own to do. However, a spatial organization of the right kind might come about for other reasons. It might arise naturally when brains are growing. Putting neurons into columns can also be useful from an information-processing point of view. A brain could end up shaped in a field-generating way without that field doing anything. In this second view, again, the synchronization of one cell with others does have effects on the brain's workings; it's only the larger electric fields that don't.

The idea that the synchronization of activity is an important part of how brains do things is increasingly influential. The brain seems to be an organ that generates and uses rhythms of activity, with these rhythms existing at many scales, one rhythm embedded in another. For a number of researchers (including Rodolfo Llinás and György Buzsáki) this view of the brain has a philosophical or semi-philosophical side to it; it supports a view of the brain as intrinsically active. The brain does not need incoming sensory information to stir it into motion. Instead, the role of sensory information is to modulate the brain's own self-generated activity. This is seen as contrasting with passive "empiricist" pictures that see the mind as merely reactive, needing to derive its patterning from elsewhere.

These ideas about rhythms do indeed alter our gestalt of what a brain is, but that is not the end of the matter. A third, more controversial possibility is that the fields themselves have a biological role. What Berger was seeing was part of the brain's operations, not just a byproduct. Despite not being a neuroscientist, I find myself making a bet on this view. I expect the question to be contested for some time, but here is how things are looking.

This next stage in the story begins with another discovery by a scientist working in unusual circumstances. Angélique Arvanitaki was a French neurophysiologist (born in Greece, as the name suggests). She worked on the nervous systems of molluscs at a marine station near Toulon, France, in the late 1930s and 1940s. Research conditions were not easy, as the Italian army occupied the marine station after the fall of France in the early part of World War II. Arvanitaki was one of those women scientists whose work appears to have been somewhat underrated—perhaps because of gender, perhaps because it seemed out of step with the main directions of the time, perhaps both (one is reminded of Barbara McClintock of "jumping genes" and Lynn Margulis of the symbiotic origin of mitochondria). A central line of work at that time was focused on the synapse, the gap between two neurons within a network, and how the gap was bridged. Arvanitaki began her most important paper, published in 1942, by saying that nerve cells could influence each other without this influence going through synapses at all, and she showed this experimentally. This sort of influence is now (based on her terminology) called the *ephaptic coupling* of nerve cells.

This finding was not directly related to Berger's waves, and work on ephaptic coupling was initially focused on local interactions between one cell and another. But recently, a connection has been shown. In some cases, the rhythmic patterns in fields generated by a whole region of the brain can affect the activity of individual cells. In particular, the timing of the activity of neurons, including their "spikes," is affected by these fields, and this appears to be one way that brains synchronize the activity of their cells. In petri dish experiments, these influences traveled between pieces of brain that had been surgically separated, but placed near to each other. As Christof Koch and Costas Anastassiou say, this is a novel feedback mechanism, "with electric fields altering the activity of the same neural elements that gave rise to them in the first place." I said earlier in this section that the flux of ions over cell borders is a bit like electrical breathing. Then: some cells breathe together in a rhythm. And now: each cell can electrically sense the chorus-breathing of the whole.

These influences are not physically mysterious. We can work through them with the aid of an analogy introduced by the Colombian neurobiologist Rodolfo Llinás, one of the central figures in the exploration of rhythmic patterns in the brain. In a book from a couple of decades ago, Llinás introduces the theme of rhythms in nature with a cicada chorus. When large numbers of cicadas sing together, a rhythmic, pulsing character can arise spontaneously. The analogy is good, but in the framework Llinás uses, there is an apparent difference between cicadas and neurons. Neurons can "hear" only neighboring neurons to which they are connected, while a cicada can hear everyone at once. In Llinás's own discussion, neighbor-to-neighbor relations are the means by which brain rhythms are established. The surprise of the newer research on the field effects described above, which came after Llinás's book, is that sometimes neurons *can* hear the activity of the whole—they are affected by the overall electric field. Brain activity seems to be even more like a cicada chorus than Llinás supposed.

During the time I was working on this chapter, I went on a summer walk at dusk in a forest. In Australia, some years are big cicada years, others less so. This was a big year. The sound made by each cicada was *breek-breek*. A slight rhythm was discernible at the level of those syllables, but the cicadas also built up repeatedly to a crescendo, the ascent taking about fifteen seconds, and then dying away a little more quickly, reaching silence about twelve seconds after the peak. They would pause, and begin to rise again. This pattern, extending a bit over thirty seconds, was quite stable.

I did some reading about why cicadas and their relatives sing. There are different possibilities and perhaps different cases, some more cooperative and others more competitive. I will introduce a hypothetical account as an illustration. Suppose each cicada is trying to be as loud and noticeable as possible, to attract a mate. Then situations will arise where one cicada sings a bit, others try to sing over it, and that gets everyone working up to a crescendo. Then fatigue begins to set in, the initial singers fall away, and so do the others. In the resulting silence, someone starts to sing and they work their way up again. Suppose that is true. Then each cicada's singing is influenced not only by its neighbors, but by the total volume of the chorus at that time. Each cicada both contributes to and responds to the chorus. The pattern is not coordinated by a conductor, but the result is a regular rhythm.

As I said above, recent experimental work suggests that the

fields generated by brain activity have effects on the functioning of the brain itself. Individual neurons are "listening" to, or at least affected by, the fields generated by a mass of neural activity combined together. This might be a minor detail, a curiosity that is unimportant to the way brains work most of the time. The fact that I am discussing it so much shows that I suspect that is not how things are, but it might be. Let's now see how all these ideas affect the questions I am exploring in this book.

One set of questions here concerns rhythms and synchronization, while another is about the fields themselves. Synchronization and rhythmic activity are fairly widely agreed to be important, while the fields are more controversial. As I noted earlier, a wide range of animals have these rhythms, including fish, flies, octopuses, and even flatworms. Did the common ancestor of all these, back in the Ediacaran, have rhythms of this kind in its nervous system? Or did similar rhythms get started independently in different evolutionary lines? They might start repeatedly at a similar pace because of the basic features of neurons themselves. Either way, the invention of the neuron was not just the origin of an excitable cell with projections to others, but of an oscillatory device, with rhythms that a great many animals have put to some kind of use.

As we speculate about the history, an even older scientific experiment, and an accidental one, becomes relevant. Christiaan Huygens was a seventeenth-century scientific polymath, in the wonderful style of that age, and the inventor of the pendulum clock. He noticed in 1665 that a pair of those clocks would naturally synchronize their oscillations, beating in time, if there was some physical link between them and their natural cycles were of similar length. Rhythmic things have, Huygens said in a letter sent to the Royal Society of London in that year, "an odd kind of sympathy" between them. As the neuroscientist Wolf Singer notes, this suggests that if you have oscillators in the brain, it may be easy to get them to naturally synchronize. Once they have done this, their rhythms can be given roles to play, functioning perhaps as various kinds of code. Some synaptic connections in the brain also appear to be designed by evolution as synchronizers and rhythm managers, consolidating these tendencies. And in the light of the recent work on ephaptic influences, one can add that once there is synchronization, if cells are also lined up in space in the right kind of way, the electric fields they generate may come to have effects of their own. They can bind the brain's activity together in additional ways, enabling it to behave as a whole.

Returning momentarily to Hans Berger, who sent us down this track, one reason he would inevitably be frustrated in a search for telepathy by this route is the fact that an electric field needs a conducting medium, and air is a bad conductor. Water, in contrast, is fine. In a way, there is not much stopping fish from engaging in electric telepathy. What's stopping them is mainly a matter of signal strength. Waves in a field generated by one fish would, presumably, be the right kind of thing to affect neurons in another fish if the signal was strong enough to get across, but the strength of this influence falls off too quickly over space, and would also probably be swamped by other activities in the "receiver" fish. Some fish do communicate with each other by means of stronger, actively generated electric fields, and fish can get by very well in any case with their long-distance touch.

What importance does all this have for questions about sentience and experience? Is this merely part of the detail of how brains get things done, or does it affect the larger questions about minds and bodies?

The element that appears to suddenly transform the problem

is the fields. They seem to give us an entirely different candidate for the physical basis of the mind. Perhaps the difficulty in seeing how conscious experience could exist in a brain has been due, in part, to the fact that we were only considering one part of what is going on there. I think there's a good deal of truth in that last thought, but some of the temptations that rush forward here are temptations to resist.

A field does look like the right sort of thing to help us with the puzzle of consciousness—a warm invisible glowing area, centered on the brain but spreading from it, a blur of energy in our heads. That blur seems to have a more experience-like shape than the rest of what has been uncovered in our skulls. But a field can't solve the problem just by being a special sort of stuff. It would have to *do* things of the right kind.

The mental image we fall into, when imagining these fields, is a sort of physicalized Rothko experience, a blurry wash of qualia. That very fact should alert us that we are falling into a relative of the fallacious moves I diagnosed back in Chapter 5, when the materialist was asked to find a way to conjure colors and sounds out of dark and silent machinery. That Chapter 5 error was cured, in part, by noting that conscious experience is the first-person point of view *of* a living system of a certain kind, not something conjured up *by* the workings of that system, something that might be seen from the outside if we knew where to look. That erroneous framework is hard to entirely discard, though, and when we find invisible fields in the brain, we seem to have been given a better material for a conjuring of color and sound, and it is tempting to try it all again. Tempting, but still a mistake.

However, I think that these discoveries do make a difference to the project of explaining experience—certainly the rhythms, and perhaps the fields. First, they change our picture of what sort of thing a brain *is*.

The way to make this clear is with the starkest contrasts. There is an old but still common picture of the brain in which it is seen as a large and complicated signaling network. For example, people used to compare a brain to an automated telephone exchange. That analogy is nearly as old as telephone exchanges themselves; the English scientist Karl Pearson made the comparison so early, in 1900, that automation was not in the picture and he had to imagine human operators manually making the connections. Later views imagined a switching network that wires itself automatically. A common view expressed by neuroscientists (I remember hearing this expressed very firmly by a Harvard neuroscientist around 2010) is that if we knew all the local signaling pathways from one neuron to another in a brain, we would have a complete understanding of what that brain does. This image of a signaling network has seemed almost inescapable to many people looking at brains, but it has often seemed a bit antithetical to explaining conscious experience. And as far as I can tell, this view of the brain is changing.

The work I have been describing here in this chapter gives us a quite different picture. A brain is a biological organ with self-generated electrical activity—rhythmic, often synchronized, modulated by the senses. Cells in the nervous system other than neurons, such as *astrocytes*, have subtle but increasingly noticeable roles. There is certainly a great deal of neuron-to-neuron signaling, but that is not all that is going on. This alternative picture of the brain is both better supported scientifically and more comprehensible as a basis for experience. If you resist the idea that the physical basis of your experience is just a signaling network like a telephone exchange, then I think you are probably right. You are a thing of another kind.

These ideas about nervous systems and brains are an important piece of the mind-body puzzle. As I see it, this is the second element encountered so far in this book that helps us bridge the apparent "gap" between mental and physical. The first element was encountered in Chapter 5, where I described subjectivity as a feature of animal selves and their dealings with the world— point of view, self and other, subjectivity and agency. The second is a view of the nervous systems that have been shaped by the evolution of agency within animals. This second element asserts the importance of what I will call *large-scale dynamic properties* of the brain. These include rhythms and patterns of synchronization, the more controversial effects of fields themselves, and probably other related activities. Some of these patterns are global—they are features of the brain as a whole—and some are more local, while still tying together many thousands of cells. The explanation of how subjective experience comes to exist, and how it can exist in a physical world, involves these factors, too. The idea is not that experience or consciousness *is* a field, in the physical sense. It's a pattern of activity to which the fields contribute, along with much else. Evolution has produced animals that engage with the world as subjects, and has also produced an organ of a biologically surprising kind that mediates these activities.

This organ is, again, not just the human or mammalian brain. A wide range of animals have these features. It is remarkable how many different animals with different brain designs have similar rhythmic patterns going on inside them. That fact brought about something of a gestalt shift for me. Some years ago, a few scientists and philosophers raised the idea that a particular kind of high-frequency wave pattern in the brain (the gamma wave) had a special connection to consciousness. This idea was introduced by Francis Crick (of DNA fame) and Christof Koch. I thought of this as something that might indeed be important in our own case—human experience—but unlikely to help with the most general questions. However, rhythms of this general kind turn out not to be a peculiarity of brains like ours; they are all over the animal kingdom. (Perhaps the researcher who has pushed this theme hardest is Bruno van Swinderen. For example, he and Ralph Greenspan found that a particular wave pattern is an indicator of attention to an object, or something like attention . . . in a fly.) From there, I realized that this work was changing in fundamental ways our view of what sort of thing brain activity is—not just in us, but in animals generally—and hence what the physical side of experience might be.

These features matter. But how do they matter? In an overall explanation of experience (sentience, consciousness . . .), what is the division of labor between the two factors on the table now—engagement with the world as a subject, and the large-scale dynamic properties of a brain? Any detailed story that might be told on this point would be full of speculation about unresolved scientific issues. But I will indicate a general shape to the situation that might make sense.

Those who emphasize the role of these large-scale dynamic patterns in perception and thought, especially the role of the rhythms, often see them as something that binds a lot of processing together, something that enables a brain to combine many different aspects of what it encounters into an overall picture or scene. This integration will have many consequences for subjective experience. At a given moment your experience might include not only a visual tableau, but also a mood, a sense of there-ness, and more. Perhaps it is not merely the bringing-together of different kinds of information that these extended dynamic patterns achieve. They might contribute to the distinctive manner in which all these goings-on are formed into an experiential whole. And perhaps in further ways they give experience, in us, some of its particular texture.

This is related to a subtle point, another one that I am not at all sure about. When we ask why seeing or hearing is (or can be) experienced, a sense of puzzlement can result from imagining what the brain is doing while the seeing or hearing is going on. We imagine neurons firing and affecting other neurons, who fire in turn, and so on. Suppose we switch to imagining this process in the way that various neurobiologists discussed in this chapter would like us to imagine it. The brain starts out full of activity, including these large-scale but integrated patterns, and the sight or the sound we encounter *modulates*—shifts, perturbs—that activity. Is it still a surprise or puzzle that *that* will feel like something?

A natural question to ask next about these two parts to my story is this: What if you, or some other being, had one without the other? In particular, what if you were an animal that had the kinds of sensing and action discussed in Chapter 5 with none of the additional dynamic properties of the brain discussed in this chapter?

Thought experiments of this kind can become a trap. We are led to think we can imagine an exact copy of a human, with the same behavior, but with entirely different physical processes underneath. However, if some person or other animal does not have the large-scale dynamic properties being discussed here, they will not perceive things and behave in the same way as something that does have them. Still, the features of subjectivity discussed in Chapter 5—sensing, acting, point of view—were, in a sense, rather schematic; it seems that an organism could have some version of them, a version different from ours, while differing in the brain's large-scale dynamic properties. This is suggested by the fact that when we work through those schematic features of subjectivity, and consider what a nervous system has to do to bring them about, neurons apparently can make all this occur while operating just as signaling pathways, relays, and the like. The basics, at least, can apparently be achieved without the particular dynamic properties discussed in this chapter. And then we can ask, using Nagel's phrase: Is there something it would be like to *be* a system of that sort? Would it undergo experience?

Well, there's *something* going on in that system, when it perceives and acts. Any system that has the schematic features of subjectivity has a lot going on inside it. Are those goings-on *experiential*? That seems the next question to ask, but what does it mean? We might be asking whether that system is undergoing a faint version of what goes on in us. The answer to that question is probably no; after all, the scenario on the table assumes big biological differences between us and them. So that question doesn't resolve the matter; we want the question to be less tied to our own case. Suppose we ask whether the organism has a point of view, whether things seem some way to it, and so on. Then the answer is yes—that is the yes we gave earlier when we said that it had the schematic features of subjectivity discussed in Chapter 5.

Can we ask something between these two questions? What comes to my mind next is a question about the sense of presence, though that sense is not necessary for experience of any kind. Many people will want to ask next about qualia. Does the being we are imagining have qualia? If these qualia have to be color fields and the like—the things people usually have in mind when they ask about qualia—then we are once again making too close a comparison to our own case. If we ask whether there is "something it's like to *be* it," then we are back where we started. Our current language seems somehow inadequate for asking the crucial questions here. But I don't think the problem can be solved just by coming up with better terms; at this point we are inside a thicket of every kind of uncertainty.

In a thicket but still pushing forward: just above I asked what it might be like to be an organism that engages with the world as a subject but lacks the large-scale dynamic properties found in our brains. What about the converse situation—something with the right large-scale dynamic properties that does not engage with the world as a subject? In a way, that is impossible, as the dynamic properties being discussed here are part of how we do the things discussed in earlier chapters. If you imagine an organism doing none of those things, it won't have the large-scale dynamic properties either. If something is just a physical system full of rhythms and electromagnetic fields—the engine of a car, for example—there's no reason to believe it has any thoughts or experiences at all. To suppose that it does is to fall back into the error of treating electrical properties as somehow mental in themselves, which they are not. Similarly, a conscious mind is not just a system whose activities are unusually closely tied together, or tied together in a way that's active and energetic. I think of that approach as seductive, but the wrong track.

The exact relationship between the factors now on the table is uncertain, though, and possibilities abound. Rodolfo Llinás, the neuroscientist who has come up several times in this chapter, has his own proposals in this area. He suggests that consciousness itself is a dream-like state, generated by rhythms, loops, and resonances in the brain's activity. Ordinary consciousness differs from dreaming in how much the senses are allowed to intrude onto, and modulate, that state. This is a view that puts more weight on the large-scale dynamic properties of the brain, less on a subject's engagement with the world. In stark contrast to Llinás is Björn Merker, a neuroscientist who influenced my thinking about the first element of my account. Merker has argued that at least some of the rhythms seen in the brain, the 40-hertz rhythms discussed by Crick, Koch, and others, are much overrated in their importance, at least on the "cognitive" side of things—thought, perception, and making sense of the world. Merker thinks, instead, that these rhythms might be important in the background upkeep of brain activity: they keep things running smoothly, preventing a descent into runaway states such as seizures. Large-scale dynamic properties of the brain might have a variety of roles in different animals and different evolutionary stages. Even when they are not doing much on the "smarter" dimensions, might they still affect how things feel?

River Riven

In Western Australia again, in whale shark territory, on a blue-water tropical dive, I was taken to a site with many deep crevices along a coral reef. Among them flowed a torrent of small, silver-green fish. There were countless thousands, coursing in a stream through the passages. They appeared purposeful, but there was no overall movement; they went back and forth on themselves, over and over. The stream looked like high-speed lava, but regularly changing course. The locals called them "glass fish," and from a distance, they did look like glass in their silver and green. But up close, there was lots of motion.

Dozens of larger fish, perhaps a hundred in all, were attacking them. These were trevally in metallic silver, some small barracuda, and others. The genius of the fish form was evident, and also its limitations. The big fish lunged in and, as they lunged, the river would part, either instantly or just before the large body arrived, so the hunter would launch into nothing. Sometimes a big fish would hurl itself in a whole-body sideways strike—*thwack*—into the stream. I think this was an attempt to stun some of the smaller fish. As the attacker arrived, the fish in the river would jolt en masse, as if a single object, and flow around it.

The trevally and the other missiles were accompanied by a couple of Rankin cod, blotchy, grumpy-looking fish with a prominent lower jaw. They hung about looking frustrated, as they lacked the agility of the trevally. But it was hard to assess whether the hunters were having any success at all, as the river went back and forth through dark, plated coral.

Such a lot of the fish way of being was on display: speed and power, bodily sensitivity through long-distance touch, and the schooling that this enables. A life-and-death acrobatic display in this small corner of the world's blue water.

8

ON LAND

Hothouse

Sun, a fierce glare of sun, meets your head as it breaks the water at the end of the dive. You emerge slowly, and as you start walking up the breakwater steps, three things are apparent: Gravity—the gear you are wearing suddenly weighs a ton. Evaporation—drops of seawater fall from you onto the dark rocks, and immediately start vanishing into the air. And the hot sun in your eyes.

Life and mind began in water, and we carry the sea with us in all our cells. But the transition to land marks its own momentous stage. Behaviors are different here; what was easy becomes hard, while some hard things become easier. The land is a hothouse, festooned with plants. The sun is the ultimate source of energy for nearly all living things, but land plants, especially flowering plants, have photosynthetic rates that outstrip anything in the sea, and the land's ecologies feature very high rates of energy flow. That sense you had, when you climbed ashore, that the sun *hit* you with energy was not an illusion. The land is full of energetic promise. If you can stay alive there, much can be done.

Leaders Once Again

The first animals to creep up into the sun were arthropods. The first of those were not insects, the group most familiar to us now, but probably relatives of spiders and millipedes. Arthropods were leaders among animals in the Cambrian, it seems, and leaders here once again.

The land is difficult, and arthropods had features that helped. Evaporation began as soon as you walked up the steps; the arthropod external skeleton can become a shield that keeps in moisture. Movement is harder on land, but arthropods were blessed with legs. Those members of their endless array of appendages gave arthropods an advantage—a leg up—for life on land.

If spider and millipede relatives began the process, they were just the start. Arthropods have made something like seven, perhaps more, separate moves onto land. A branch of the arthropods now called "Pancrustacea" brings together the familiar crustaceans of Chapter 4 with various others. Members of this group alone came up onto land several times, and one sortie (whose timing is controversial) gave rise to insects. In this new home they underwent enormous, limitless diversification. A large majority of all known animal species today are from this one group, the insects.

Insects are mostly land specialists, and a large part of their story is their coevolution with land plants. This association between two huge groups features cooperation and conflict in many different forms. It is somehow tempting to think of plants as old, but they are not as old as many of the animals I have been discussing here. They originate in green algae, one of the many ancient seaweed-like organisms, and had their own transition to multicellular life. Probably fairly soon after the Cambrian, some modified algae forms began to make their way onto land, and by the Carboniferous, around 350 million years ago, plants were visible as sturdy green structures. At that stage they included ferns, cycads, and conifers, and later, by the time of the dinosaurs, flowering plants.

Land plants, especially flowering plants, consume the sun's energy with an intensity and efficiency that far surpasses sea life. Insects became tightly knit into this energy flow, not just as passengers, but as part of the process by which land plants evolved their special features. Insects consume plants, but also became the pollinators that made flowering plants possible.

The lives of insects in this demanding but rich environment have produced some unusual modes of existence. Arthropods of the sea feature elaborate innovations in and over their bodies. Arthropods on land, while not lacking unusual bodies, have explored wild innovations in their life cycles. In some insects (tiny parasitic mites), the eggs hatch while inside the mother, yielding a brood of sisters and one male. The male mates with the females, the mother dies, and the now-pregnant females eat their way out. In scale insects, the females are the closest thing dry land has to a stationary, never-moving animal form, while the males fly, never eat, and live for a couple of days. These and other eccentricities are ways to achieve massive fecundity and make maximum use of ephemeral food sources. Insect life features vast numbers of near-immediate deaths.

Though many insects are miracles of miniaturization, they show feats of complex behavior and cognition as well. Bees are especially remarkable cases. In learning, they are able to deal with abstract relationships in ways that are very unusual in animals. Some of these achievements are so intricate that it is hard to describe them briefly. In a 2019 study, bees had to learn that if they entered a chamber and initially saw a yellow display with (say) three items, the rule to apply at the next decision point was to choose a display that had one more item (four, in this case), but if they saw a blue patch, they had to choose a display that had one fewer item (two). They learned this pretty well.

Like the fish of the previous chapter, bees can learn by imitating the behaviors of others. And when bumblebees have learned to prefer a spatial arrangement that is presented in scent, they can extrapolate this and choose the same arrangement when it is presented visually. Bees, in their well-organized vegetarian way, show what insects, often much scruffier, can be capable of.

Bees reserve perhaps their most impressive and also beautiful behaviors for projects of physical construction. This was shown in some very old experiments, from around 1814, by François Huber on the making of honeycombs. Bees build honeycombs between surfaces the wax can stick to. They find glass to be a problem surface and avoid it (Huber noticed that they didn't build combs on the glass viewing panels he put into hives to watch his bees). When confronted with glass, they respond by constructing unusual shapes in their combs. In one experiment, Huber put a glass panel into a wall after the bees had started building toward it, but before they were very close. The bees altered their path and curved the comb gracefully around to meet a wooden surface.

Bees are sometimes compared to octopuses, in a sort of cognitive contest between invertebrates. Bees have much smaller brains—one cubic millimeter—but pack a great deal of complexity into that space. Any talk of a contest makes little sense here, though. The two have different lives and bodies, and very different styles when confronted with behavioral problems and tasks. Bees are masters (mistresses, more accurately) of logical abstraction and intricate physical construction. They have highly organized lives with specific roles. A bee researcher once told me that one reason so much is known about bee behavior is that they will do exactly what you need them to do, over and over, repeating the behaviors just as the editor of a scientific journal would like. They do scout new options among all this (as another bee researcher added), and that is part of their smartness. But they are very task-focused, and that makes sense given how bees live, securing honey and maintaining their colony. An octopus is a very different sort of beast. A few years ago, Jonas Richter and collaborators presented octopuses with a puzzle box, a task in which octopuses had to learn to produce two actions in a sequence (push then pull). Some of them did learn to do this—the experiment was a success—but the successful behaviors were embedded in a huge amount of random manipulation. They got better at producing the right sequence, amid a lot of chaotic wrangling. I am not sure how the logical side of this task relates to those that the bees have mastered, but they each mastered their tasks in an entirely different way: the bees, through internal computation; the octopuses, through exploratory manipulation.

Bee behaviors, by animal standards, show relatively little noise or disorder; octopus behaviors have little *but* noise and disorder, but somewhere in there they get things done. The octopus mode of behavior reminds me of Ted Hughes's poem "Wodwo," about an imaginary creature trying to make sense of its existence and its own actions. Hughes ends up inside what seems to me a kind of terrestrial octopus, an engaged but somewhat unhinged explorer. (". . . I've no threads / fastening me to anything I can go anywhere / I seem to have been given the freedom / of this place what am I then? . . .")

Sense, Pain, Emotion

Insects are the most numerous animals on land or anywhere else. How do they fit into the story of the evolution of experience?

The question of a conscious insect is difficult and also rather fraught. It would be, for many people, a considerable gestalt shift to think there is *anything* felt, anything experiential, going on there. The shift would also be unsettling, given the carnage of pest control and the carnage seen in ordinary insect life. This is a place where the term "conscious" also seems a problem. That is how people usually consider the issue: Are insects conscious? But all we are asking, at least at first, is whether there is felt experience of some kind going on, perhaps in a minimal form.

Earlier in this book I looked at crustaceans such as crabs and shrimp. Some crustaceans probably do have experience. Insects come out of the same branch of the tree—they are an evolutionary outgrowth from crustaceans, one mostly suited to land. They also have a fairly similar design. But those relationships do not settle the matter; insects have gone down their own road.

Many insects are impressive on the sensory side of things; they have good vision and can fly. Flight is a behavior featuring especially complex feedback between action and the senses, the sort of feedback that contributes to a point of view. Let's now turn to another aspect of experience: pain, pleasure, and feelings of that kind. Though philosophers obsess about vision, pain and pleasure are probably what come to mind when people not steeped in recent philosophy think about basic forms of experience in animals. How do insects look on this side?

The picture is very different. Some decades ago, Craig Eisemann and a group of colleagues at the University of Queensland in Australia argued that insects do not feel pain, as all known insects appear completely unconcerned about damage to their bodies of even the most severe kinds. Wound tending has never been seen in an insect. After injury these animals just continue, as best they can, with whatever they have to do. They may show some initial squirming, but then they get back to work.

One response to this finding is to gestalt-switch back again, and conclude that insects are outside the realm of experience altogether. But that would be too fast. There is also the possibility of a separation between different forms of experience, which I will call *sensory* and *evaluative* experience. (They might also be called sensory and evaluative consciousness.)

The sensory side is related to perception, point of view, and the registration of what is going on. The evaluative side is related to pain, pleasure, and the marking of events as good or bad. Both can be described as involving how things *seem* to an animal; things can seem to be a certain way on the sensory side (it's getting colder . . .) and also on the evaluative side (. . . and that is not good). We humans have experience featuring both of these elements, and so, most likely, do various other animals. But perhaps the two can come apart in some cases. Might it be possible to have one and not the other? In this book, I have lumped the sensory and evaluative sides together much of the time. But now we should look more closely.

The idea of distinct sensory and evaluative sides to experience has consequences for philosophical puzzles in this area. I looked in an earlier chapter at thought-experiments where we imagine an organism as a physical being but imagine subjective experience "switched off." Some people think the ease and naturalness of this thought-experiment shows there is something wrong with materialism. Whether or not it does, with the distinction between sensory and evaluative aspects of experience in hand, new questions appear. Might we turn off either side of experience and leave the other intact? It seems easy to turn off the evaluative side, leaving the animal with a flat or robotic quality. Even that is not entirely clear, as flatness might be an evaluation of its own. Still, I think we can conceive the possibility of an *absence* of evaluation in a creature that has sensory experience. Perhaps it's not quite so easy on the other side? When a person imagines an insect or some other animal as unconscious but doing complex things, are they really imagining the sensory side to be blank? When we watch complex robot characters in movies, we can, in a paranoid mode, imagine a disturbing flatness on the evaluative side, but the way we imagine those beings often features rich sensory experience, with a definite point of view.

Those differences between how we imagine sensory and evaluative experience might be seen as pointing us toward a further philosophical conclusion. Perhaps the real difficulties in this area are found on the evaluative side—*how can a physical system feel pleasure or pain?*—and the sensory side is not a big deal. Perhaps given that bacteria and sponges do sense their environments, they must have sensory experience—there are ways things seem to them, and so on—but that doesn't show that they can feel things to be good or bad. This move would radically transform the whole problem, but I think it would be a mistake. Cameras, phones, and thermostats also "sense" things, but that does not show that they have any kind of experience. The sensory side, as well as the evaluative side of experience, does raise mind-body puzzles.

Let's take a closer look at insects and others with this distinction in hand. Back in Chapter 4 I introduced some ideas about pain. Nociception is the detection of damage, tied to a behavioral response. It is very common in animals, but can lend itself to an interpretation as mere reflex. As a result, people look for markers of something more, markers of something that might be tied to feeling. The options include: (1) tending and protecting injuries, (2) seeking out analgesic chemicals, (3) learning to avoid particular behaviors or situations, and (4) trade-offs in choice situations, where an animal seems to balance the badness of an experience with other costs and benefits. (We looked at trade-offs made by crabs in that earlier chapter.) These are all seen as behavioral tests for the presence of felt pain, or something like it.

When we look again at insects, with tests like this in mind, we find that one test the insects do pass is the one involving learning. In particular, some insects can learn to avoid situations where they will be exposed to excessive heat. But insects have still never been observed tending and grooming injuries; that claim from the old no-pain paper still holds up. Crustaceans, as we saw earlier, do things like this and so do octopuses. Julia Groening and her colleagues (also at the University of Queensland) decided to see whether bees who had been slightly injured would seek out analgesic drugs. Tests of this kind have been applied to chickens and other animals, and can be quite convincing evidence for pain. It is then worth asking whether bees, who are very smart animals, will seek out morphine when they are injured with a pinching clamp or the amputation of a limb. Groening and her colleagues found that they do not. Reading between the lines, I think Groening and her team were a little surprised at this result.

That result takes us further toward a view in which insects have rich sensory experience but much less, or a complete blank, on the evaluative side. But against that background we then encounter another body of work. If our interest is evaluative experience, pain and pleasure are not the only things we might look for; other candidates include *emotions* and *moods*. These are not acute and specific, as pain is (or can be), but longer-lasting states that affect all sorts of other decisions. Such states include fear and anxiety, and also their more positive analogues, though the negative ones have been the main subject of research. These have now been seen rather clearly in insects, and also in some other invertebrates. This work opens up a new way of looking at the evaluative side of animal experience.

Terry Walters of the University of Texas has worked for many years on a fear-like state called *nociceptive sensitization*. This is a heightened sensitivity after damage that alters the animal's response to various *other* choices and stimuli. The animal seems to have an overall wariness, a caution that can last for hours, days, or weeks, depending on the case. A similar emotion-like state was seen by Melissa Bateson and her colleagues in honeybees. They found that if a bee was shaken, it induced a kind of pessimism: the bee became inclined to treat ambiguous stimuli (midway between what had been good in earlier experience and what had been bad) in a pessimistic manner, assuming the worst. On the positive side, bees can also be induced to be upbeat or optimistic. Cwyn Solvi and her collaborators in Lars Chittka's laboratory showed that an unexpected reward has an effect on bumblebees that is the flip side of the pessimistic mood that Bateson's experiment produced. Good moods are also seen in fish.

This work is rather convincing. The animals' responses are nothing like reflexes; they are integrated into all sorts of things the animal is doing. The animals seem to enter a kind of pervasive positivity or negativity. Some other discussions of evaluative experience are organized around a distinction between immediate responses, which might show acute pain but might be reflexes, and learning, which seems to be a sign of something more sophisticated. These emotion-like states are in between those two timescales—longer than a jolt of pain, shorter than learning. When we look at these intermediate scales, the evidence for experience can be quite strong.

What should we make of all this? One possible conclusion is that the old no-pain paper by Eisemann and his colleagues was wrong, and insect pain or something very like it is real and extensive, though initially hidden. Another conclusion is that insects do not have pain-like experiences of damage, but do have something more like moods and emotions. Evolution might have suppressed and modified some features of experience in insects, to fit their short, routinized lives. The distinction between pain and emotion that I have been assuming here might also be questioned. One recent paper investigating emotion-like states in flies argues that what we find in the animals looks like what in humans might be seen as chronic pain. Certainly we should not just import the familiar human category of "pain" into insect life; emotions, moods, and pain might all shade into each other in novel ways. Perhaps insects inhabit their bodies like vehicles (a lost leg is like a flat tire or broken windshield), but they can feel a kind of overall malaise when things are going badly, and this feeling affects their decisions.

In trying to work out even the basics of what is going on, no end is seen to the puzzling hints and intriguing observations. In Julia Groening's study of bees and morphine, the bees did not take morphine when tight clips were put on their legs, but they did try to get the leg clip off in a very recognizable way. "We observed, however, that some individuals were stepping on their clip and pushing it down with another foot, presumably in an attempt to remove it." This observation goes by in a moment in their paper, but somehow it seems quite telling. It invites an imaginative projection into their little minds. I assume that a leg clip would be a novel problem for a bee—perhaps not; perhaps if something sticks naturally to a bee's leg with resin, the same thing is done. But if this is an improvised behavior, it is a rather human-like one, and some sort of glimpse into the being of a bee.

Varieties

Emotion-like states are seen both in insects and in another group that has wandered onto these pages several times: gastropods. These are slugs and snails (including the Gaudí-like nudibranch

Tritonia of Chapter 3). These animals are where work on emotions and moods in invertebrates began.

Gastropods are molluscs, like octopuses, but they tend to have nervous systems with only tens of thousands of neurons, rather than the million of a bee or half billion of an octopus. They also have much simpler eyes than those animals. Gastropods do have a sophisticated sense of taste or smell. I have watched nudibranchs in the sea following long paths toward each other—very slow paths, but very direct, despite rough terrain and buffeting currents.

Insects, as we saw, have good senses and enigmatic evaluation. Gastropods might have the opposite combination. A scene that illustrates this possibility in gastropods was captured on video by a diver, Steve Winkworth, who frequents some of the same spots that I do. A "bubble shell" is partway between slug and snail, with a small but colorful shell, and a spectacularly folded and luminous body. This animal was moving over the reef and it found itself in a field of skeleton shrimp, the tiny sharp-taloned arthropods that lurked in the bryozoan colony in Chapter 4. Those animals occasionally form dense clusters or carpets in a particular area, and that is where the bubble shell ended up, in a field of flailing, pinching attackers. As the skeleton shrimp began striking it with their claws, the bubble shell flinched, turned in what seemed like alarm, and tried awkwardly to make its way out of the field. I suspect the bubble shell had no idea what was going on—its eyes were nowhere near up to that task—but whatever was going on, it was not good.

A couple of times in the shallows of the same waters, I've come across giant *Aplysia*. These are sea slugs, but on a different scale from other gastropods. The largest I've seen was well over two feet long (70 cm). When they are active, they have a very unusual way of moving, reminiscent—of all things—of the gallop of a horse, with front down and haunches raised, then front raised and rear down. As the "front" in this case is the animal's face, this is no ordinary gallop. With two small wing-like protrusions on the back, the result is a molluscan Pegasus. I knew this animal had a small nervous system, but as it galloped along in front of me (on its face), I found it impossible not to attribute some sort of experience. I was reminded of a comment by Daniel Dennett, who noted that when we watch animals far from ourselves, we are enormously influenced by their "behavioral tempo and rhythm." If an animal moves extremely slowly, or in a very clunky way, it tends not to cross our minds to see it as a conscious being. Once animals start moving at *our* pace, with human-scale responsiveness and deliberateness in their motions, we see them differently. The *Aplysia* had a nervous system similar to that of countless tiny slugs I had seen in the water. Those other encounters included little temptation to put oneself into their gastropod shoes. But all one has to do is scale them up to giant *Aplysia* size and have them move at a gallop rather than a slow crawl, and suddenly experience in these animals seems almost inescapable, or at least far more feasible.

We are beginning to get a picture here of *varieties of subjectivity*, different ways of being a subject that relate to an animal's lifestyle and circumstances. An animal might have great complexity on the sensory side—flying, chasing, landing—but have goals so simple and definite that very little processing of what to do next is ever needed, and damage is an unfelt inconvenience, something like a flat tire. In insects, the adult stage is often "a disposable reproductive machine," as the zoologist Andrew Barron has put it. As insect bodies are difficult to heal, and so much in insect life has to be done on a tight schedule, there will often, as Barron again says, be no point in trying to protect or tend an injured area: one should just "soldier on." In cases like these, insects would have little use for acute feelings of pain; such feelings could do them no good. If the package of features insects inherited from their crustacean ancestors included the capacity for pain, evolution on land might slowly rub this out.

As Terry Walters notes, gastropods have more apparent use for pain-related behaviors and, I would add, for feelings that go along with them—Walters is cautious about the "felt" side. Gastropods have soft bodies that are sensitive but able to heal well, and their life spans are often around one to two years. A gastropod may well improve its lot, if it has initially been damaged in some way, by avoiding further injury and giving its body a chance to heal. For marine crustaceans, whose lives can extend for not just years but decades, whose schedules have none of the just-in-time regimentation of many insects, pain also makes more sense.

Once we take on board the idea of deep differences between varieties of subjectivity, insects and gastropods are not the only interesting cases. Sharks and rays, unlike other fish, seem to lack nociceptors, and some of their behavior suggests that they cannot feel pain. They appear surprisingly oblivious to stingray barbs, for example. As Michael Tye notes, in this way sharks may be akin to insects. Again as with insects, sharks do have other behaviors related to evaluation and feeling, including learning by reinforcement. The reef sharks at the cleaning station described in the previous chapter gave an apparent shudder at being nipped by small fish—they did not seem oblivious. And Culum Brown of Macquarie University, the expert I consult on shark-related matters, does not believe the no-pain view of sharks at all.

Bony fish, such as trout, show extensive evidence for pain and also for pleasure. At cleaning stations, some cleaners use their fins to stimulate their client fish in a kind of massage. The massage does not remove parasites or bring other obvious benefits. A nice (in both senses) experiment by Marta Soares and her collaborators found that in an artificial cleaner-fish arrangement, if a fish was "cleaned" by a moving model that included the massage, it had lower stress hormone levels than an unmassaged fish.

In the course of this chapter, we've encountered some animals with nervous systems so small that it is tempting to just rule them out as potential subjects of experience. My discussion seems to be working its way through simpler and simpler animals. Is there a point where we need to just draw a line? My answer is: certainly not yet. I am influenced here by the case of crabs and other crustaceans. They had been thought by many to not even be "in the game" as far as experience is concerned, but we underestimated them. We may also be underestimating gastropods and many others. Gastropods are often chosen as experimental animals because they are simple, easy to work with, and don't trigger much attention from research ethics committees. But some researchers have started to express unease about the situation. As Robyn Crook and Terry Walters note in a review article about molluscs, it is because we think these animals have relevant similarities with us that time and money is spent doing experiments on them. But the more similarities there are between them and us with respect to pain and its relatives, the more questionable it becomes that we choose to perform these experiments. Crook and Walters—who are both respected and mainstream researchers, not outsiders or critics of the whole enterprise—end up calling for more care and oversight, the use of anesthetics, and a reduction in the number of animals used. Forty years ago, saying this about slugs would probably have been seen as ridiculous.

As with the experiments on crabs done by Elwood, the experiments done so far on pain and suffering in molluscs may well end up doing considerable good from the point of view of the animals themselves. Without these experiments, people might well have continued mistreating them in large numbers with hardly a thought. Now, their felt experience is at least a topic of reflection, and some moves are afoot to change what we do. For these reasons, I am glad the Elwood work was done. A small number of crabs were sufficient for that work, too, and their treatment was not very bad. For the same reasons, I am cautiously glad of the mollusc work, though it is often more destructive to the animals. In the mollusc case, the research is also driven in large part by the goal of understanding human pain, and that introduces a different accounting into the situation. Certainly, along with Crook and Walters, I would like to see more consideration for these animals in experimental work.

After seeing marks of sentience in insects, snails, and the like, it is always possible to say: "Fine, but can they do *this*?"—setting a new requirement or qualification, discounting the earlier ones. Asking further questions is entirely reasonable, but it is important to keep in mind how far we've come. I suspect that most of us begin by looking at bees and flies, for example, as little flying robots, and at slugs as amorphous non-subjects. The idea that they have emotion-like states, in particular, is surprising against that background. And that is where we now are; we have come a considerable distance. This does not mean that we should undertake a 180-degree switch and decide that insects and slugs are just like us, and hence should be treated just like us. We need to take seriously the idea of varieties of subjectivity, and these varieties might bring with them different practical and ethical consequences.

Plant Life

Plants have been an important part of the background in this chapter. Let's bring them out of the shadows and look at them in their own right.

The energy that powers life and its building of living matter comes almost entirely from the sun, harnessed by photosynthesis. In the sea, photosynthesis is seen in the bacteria who invented it, other microbes and algae who internalized the bacteria, and various others who corral and collaborate with them. Eventually, some green algae made their way from a freshwater life onto land, perhaps in the Ordovician, and grew in time into mosses, ferns, and huge trees.

Plants and animals have similar resources at the level of the cell, with the exception of the bacterial remnant that enables plants to photosynthesize. Although their cells have much in common, plants set out on their own evolutionary path, one of stillness and growth, drinking water, air, and light, reaching simultaneously up to catch energy from the sun and down to moisture below.

Might things be done entirely differently? Might there have been a mobile crawling sunbather who seeks out light and water? A life of this kind might be approached from several different paths. A few organisms have made their way some distance toward it from an animal point of departure. Various animals can photosynthesize by forming partnerships with algae or purloining their parts, and though most of them (such as corals) live fixed in place, a few can move. These include some relatives of the marine slugs of Chapter 3. All those animals live in the sea, and solar power seems to be an accessory in these cases, not their main source of energy. No one has gone down such a road from a starting point as a plant. The closest case I know is *Volvox*. These are green algae found in fresh water, with some species living in mobile colonies like tiny spherical spaceships that swim toward light. Mixed lifestyles combining movement and photosynthesis are uncommon in multicellular organisms. Once you are on a photosynthetic path on land, with its difficulties of motion, more seems to be gained from staying in place and building light-harvesting towers.

Plant cells have the means for sensing and response, and can even have action potentials, but plants do not have nervous systems in anything like the usual sense. Within this slower and more physically constrained life, they do have their active side. A few, by transporting water and varying the stiffness of their parts, can move them visibly. The best-known examples are the Venus flytraps, a small group of plants that trap and eat insects. Those cases are rare, and if you discuss plant behavior with someone who takes the plants' side, they will be at pains to convince you that *growth* also counts as an action, a slow but real one. So is making chemicals. Those are the ways plants do most of what they do.

If you also ask plant people which plants are the smartest, the usual reply is climbing plants and vines. In plants, a lot of the action takes place under the ground, in the exploratory probing of roots—Darwin realized this; he said that down here, not up in the sunlight, is where one might find a kind of plant "brain." Above ground, climbing plants have to make many more decisions than other plants, and it is no accident that they feature in experiments and surprising time-lapse videos. Climbers and vines are later arrivals in plant evolution. Almost all are flowering plants, a metabolically high-powered group that arose in the age of the dinosaurs and replaced conifer-type plants, or gymnosperms, in many habitats. Climbers seem especially awake as plants; they seem to rediscover active capacities latent in their cells, capacities that had been submerged for a time in quieter ancestral forms.

Inside a plant is also a great deal of chemical signaling. I will describe an example that surprised me. A 2018 study found that when a leaf of a plant called *Arabidopsis*—a small weedy plant that is the main "model organism" in botany—is chewed by a caterpillar or torn by an experimenter, a signal is quickly sent to neighboring leaves, prompting them to prepare a chemical defense. This signal is sent using a chain reaction of events with a considerable similarity to such processes in animals. Glutamate, a widely used signaling chemical, is released from one cell and affects ion channels in others. As a result, an undamaged leaf some distance away can prepare itself within minutes after damage to another leaf.

How do plants fit into the story that is developing about experience? Drawing on our exploration of animals, two themes are on the table so far: the shaping of living activity into a *self* with a perspective, and the integrated nervous system activity underlying those selves. If these are what matters, plants do not have much of what matters. They do sense and respond, and are full of signals, but this is probably not enough to have felt experience even in simple forms.

This is due in part to what sort of thing a plant *is*. Back in Chapter 4, I described *modular* organisms, made up of many repeated and partly independent parts, and I compared these to more integrated (or "unitary") organisms like us. The modular design is a road taken by plants, fungi, and some animals. A modular organism has less clear individuality; it is partly a community or colony, rather than an individual organism.

In the case of plants, this was recognized at least from the time of the German poet and naturalist Goethe, and Erasmus Darwin, Charles Darwin's grandfather, in the late eighteenth century. In an oak tree, or a similar plant, a module that includes a single stem segment, leaf, and bud is in some ways the basic unit, and the tree is an organized colony of these. This view can be readily motivated by experience with cuttings and regeneration, and it turns out to be true of the workings of plants in a more general way—though roots, interestingly, are not modular in the way that aboveground parts are.

A plant is in some ways like a community rather than a single individual. Within a community there can be signaling and coordination; you and I might communicate closely, and still be two. Social interaction does not merge our subjectivities. In special cases, connections might get so tight that communication between two agents dissolves into interaction within a new whole. But the mere presence of signaling does not create a fused self.

Plants I said are in *some* ways like communities; the divides between the units are less clear than they are in some other modular organisms, like corals. But a plant has less of a *self* of the kind seen in the animals I have been following in the last few chapters. Subjectivity is about engaging with the world as a self, and having things seem a certain way *to you*. Plants have less of a *you*. In some ways, a plant is a *they*, a they of stems and shoots, with signaling running between them. But that is a simplification, too; a plant is partly a community and partly an individual.

Working in a garden, when you cut or pull some of a plant

away, slicing at a stem, you can think of this as pulling one living unit away from a collection of them, and you can also think of the whole as a single organism that you are removing a piece of. With plants, I think it is right to go back and forth through this gestalt switch. Plants are not tied together like animals, our usual model for thinking about whole organisms—I don't agree with the view that plants are just "very slow animals," as the biologist Jack Schultz has put it. But plants are more tied together than a colony of distinct living things. Plant life has, on its path alongside animals, developed a different kind of organization. As a plant is less of a self, the problem with plant experience is not just the lack of a nervous system, but a difference in the kind of being that a plant is. With plants, some agency exists at the level of the cell, some at the stem or module, some at the whole physical bush or tree, and, in some cases, some at the level of the clone—a stand of plants that came from one seed and are connected by their roots.

As plants lack nervous systems, they also lack the large-scale electrical patterns that a nervous system generates. Some caution is appropriate here, as plants do have a wealth of electrical activity, new forms of which are steadily uncovered. Further electro-botanical surprises may be waiting. But merely finding *something* electrical in plants does not show that they have experience. In us, the electrical patterning in our brains integrates and forms a state of activity that is modified and perturbed by the events we sense. To show that plants have feelings, an electro-botanical surprise would have to have this form.

Something about the idea of plant experience is so striking that even slender evidence makes the issue come to life. The vision of trees—huge, still, quiet—as experiencing subjects, processing at their own pace while we scurry around them, is such an evocative one. I lived for a while in a redwood forest in California. The forest had been logged in the nineteenth century to build San Francisco houses, and most of the trees around me were just a hundred years or so old. A few, never cut, were lone giants with ages measured in thousands of years. Though I knew that I was not looking at individuals unified in the way animals are, the idea of these beings witnessing the centuries could be almost overwhelming. In the experiment I mentioned that found leaf-to-leaf warnings of damage, the sensing of damage and signaling are simple by animal standards, but most of us probably didn't expect anything like this to be going on at all. A door is opened to seeing plants differently. But if we do want to know about felt experience, it is not enough to show that plants sense and respond to things going on around them. Single-celled organisms do that, too. A plant, with its countless cells and all their chemical brewing, can have more of this going on than a bacterium or protozoan does, and "more" can make a big difference in biology. But it has to be the right kind of "more."

As work on sensing and signaling in organisms like plants has progressed, the term "minimal cognition" has been introduced to describe what goes on inside them. Minimal cognition is a package of abilities, including sensing and responding, perhaps also bringing the present and past together in working out what to do (as plants and bacteria can), all in a way that reflects the importance of the information to the vital projects of the organism. At present, it appears that all cellular life can do some of this, including fungi, plants, and single-celled creatures. Minimal cognition turns out to be closely tied to the to-and-fro traffic that is essential to life itself.

The idea of minimal cognition is not empty—it doesn't apply to everything. Compare the response of a plant's roots to water with the response made by a teaspoon of salt. The roots change their direction of growth; the salt dissolves. Both of them change—both "respond," in a sense. But the response of the plant is more than something that just happens. It is also a change in accordance with the role that water has for the plant's vital projects, for its continued existence and reproduction. A pathway has been built in the plant—with hormones and genes involved—that brings it about that the detection of water has this particular effect. The teaspoon of salt does not engage in minimal cognition.

In a sense, anything with minimal cognition has a "point of view." Given that, does minimal cognition imply a kind of minimal sentience, minimal experience? Do the two go together, so that as cognition fades away to the simplest forms, so does experience? I see the appeal of this view, but I think it is too simple. Some kinds of minimal cognition can occur without much of a self being present, and in those cases there isn't the right setup for genuine subjectivity. A plant might be a user of sensory information, a being attuned to events and responding to them, without any form of felt experience.

It is hard to even think about the borderlands, the minimal cases, the beings that have the faintest hint of something *like* experience. A view like mine implies that there must be cases of this kind. Someone inhabits the borderlands, and the question is who we will find there. Perhaps it's plants, but I doubt it. Not because what they do is simple compared to animals, but because they are on a different road.

Plants are an alternative way of putting the resources inherent in complex cells to work, an alternative that results in some sensitivity and smartness, but not felt experience. I am not at all certain about this absence of experience, especially as plant surprises continue to come in. I said about insects: "Look how far we have come—who would have thought anything remotely like emotions would be found there?" We are not yet in a place like that with plants, but we have come some way in their case, too.

9

FINS, LEGS, WINGS

Difficult Times

At some stage in this journey, perhaps around 380 million years ago, a different group of animals came up on deck: vertebrates, our group.

In the case of arthropods, several groups made their way to a wholly terrestrial life. The vertebrate story was different. A simple picture has it that vertebrates made just one such move, one momentous step. Some fish, from an old group called "lobe-finned fishes," came up and gave rise to land vertebrates of many kinds. A fuller story has not one move but many, with most of them partial—sorties into borderlands—plus one that went deeper. Those adventurous fish initiated an ongoing radiation of terrestrial animals, including mammals and birds.

This shift was hard for vertebrates. Arthropods, beginning with an abundance of legs and an external shield, had a fairly obvious road. In contrast, a torpedo with fins seems among the least likely organisms to make it on land.

Dry land for vertebrates is, indeed, a litany of obstructions.

Moving under the weight of gravity is one. Feeding was probably difficult in many ways, including an unexpected problem: swallowing. When a fish has seized food, swallowing is easily achieved by sucking the food and accompanying water into the stomach. But this relies on the similar densities of prey and water. On dry land, that sucking motion will bring in a lot of air and leave the food where it is. The invention of the jaw only goes so far. This difficulty is illustrated by the eel catfish, who often feeds by seizing food on land and then returning to water to swallow it.

The most obvious change that permitted vertebrates to achieve a permanent life on land was a change in the body form: the evolution of the "tetrapod" body, with four limbs like ours. This body is not one we should picture arising as fish try to crawl up a beach. It arose in water; it was a body for crawling and clambering in plant-choked streams. Lungs seem another obvious terrestrial requirement, but lung-like pouches had been part of some fish for a long time, used for buoyancy and sometimes for breathing. According to Jennifer Clack, whose 2012 book *Gaining Ground* I draw on often in this section, many fish already had lungs before any spent time on land.

Early lobe-finned fish seem to have been shallow-water specialists. The lobe-finned fish that remain alive now (not counting ourselves) are a few lungfish (very much in the shallows) and the coelacanths, who have moved back to the deeper sea. Video from submarines suggest that coelacanths are quite social animals, resting in groups in caves. According to Jennifer Clack, each coelacanth individual has a unique pattern of spots, and the groups remain stable in a way that suggests that individuals recognize each other.

Another challenge for vertebrates on land was the unsuitability of their eggs. After a period in which all land vertebrates lived amphibiously, one group—the *amniotes*—evolved an egg that provides their embryos with a "miniature pond." This enabled other ties to water to be loosened. Soon after that came another evolutionary branching—insignificant at the time, as they all are. An early amniote species split into two lines. From there, two branches of vertebrates grew out and radiated. Initially, a group called the Synapsids was larger, more numerous, more diverse. But the Synapsids suffered dreadfully in a colossal mass extinction, in which a majority of all animal species died out. The other vertebrate group, who had been lurking in less conspicuous form, took over. That group, the Sauropsids, included the dinosaurs.

Right: before the more famous mass extinction that killed the dinosaurs and opened the way for mammals, an earlier extinction jolted and re-sorted land animals. Prior to that extinction, which occurred about 252 million years ago, animals that were close to mammals, in both genealogy and appearance, were fairly dominant. Replete with herbivores and carnivores, they included species of many sizes, some on the scale of large bears. The mass extinction that hit them may have been due to a combination of causes—volcanic activity, meteor impacts—all of which transformed the climate. Over two-thirds of land-dwelling animal species disappeared, and a much higher fraction in the sea. Emerging from the rubble, a formerly less conspicuous group dramatically diversified, producing the dinosaurs.

The body form of the first known dinosaurs included an upright pose and grasping hands. Dinosaurs and the group including mammals reached this kind of body independently. During the earlier stages of vertebrate life on land, animals confronted the world with more of a sprawl—legs splayed, body horizontal—as seen in salamanders and crocodiles now. The transition that had to be made was to hoist the body up over the legs, whether one then goes on to walk on four legs or on two. Dinosaurs seem to have been hoisted up in this way from the beginning, and the earliest were also leaning back on two legs, with arms and head freed to engage objects.

Once given free rein, dinosaurs explored a great diversity of sizes and forms, until their own mass extinction, right at the peak of their dominance, in the age of *Tyrannosaurus rex*, about 66 million years ago. The extinction this time was due to a single devastating, climate-wrecking asteroid impact.

In the earlier extinction that opened the way for dinosaurs, large mammal-relatives were lost but some smaller ones made it through. In the second extinction, almost every dinosaur of any size was lost. Mammals themselves arose in the first period of dinosaur rule, the Triassic. They remained small all through the dinosaur era—none bigger than a badger—and a few of them made it through the extinction at the end. Of the dinosaurs, just one group survived, a group that had turned wings and feathers originally evolved for different purposes into means for flight. The single branch of the dinosaurs that remained alive was the birds.

Emerging from this latest round of wreckage, mammals expanded into a great range of lifestyles. An early mammal from this time, known from fossils dating from just a couple of million years after the dinosaur extinction, is *Torrejonia wilsoni*. It was small, tree-dwelling, with long arms and legs; an animal all warm-blooded clambering agility, blinking: a primate.

In earlier chapters of this book, suspended in water, we saw a succession of marine bodies that mark stages in the evolution of action. Cnidarians show the muscular organization of movement on a new scale. Arthropods evolved novel forms of mobility and manipulation, guided by acute senses, but with a kind of restriction or fixity in what they do. Octopus bodies, in contrast, are as behaviorally "open" as bodies can be. An octopus can readily manipulate objects that no octopus has ever manipulated before. Cephalopods also evolved large nervous systems, but on a decentralized design. Vertebrates are more centralized, but when they were fish, there was not much they could *do*. Evolution in the sea, then, produced a capacity for manipulation, openness of bodily action, and centralized braininess, but no sea animal evolved with all of these at once. That is the combination we see at last in land vertebrates, especially from the Triassic onward—a trio of features finally brought together. This combination arose independently in the two big branches, in early dinosaurs and mammals. It was transformed again in the dinosaurs who survived—birds—and came to a particular fruition in primates like us.

Our Branch of the Tree

Several chapters ago, we tried to glimpse the common ancestor of ourselves and octopuses (also ourselves and bees), and imagined a flatworm-like creature in the sea, perhaps 600 million years ago. The common ancestor of ourselves and birds is a much more recognizable relative, with four legs and good eyes, living a terrestrial life. A stumpy-looking lizard might be a reasonable thing to imagine, clambering through a swamp a bit over 300 million years ago. As we saw, the two sides of the subsequent

evolutionary branching into Synapsids (our side) and Sauropsids (the dinosaur side) alternated in how well they did during the chaotic years that followed. In evolutionary terms, they went through some changes in parallel, acquiring new similarities independently, and in other respects stayed very different. Upright, manipulative bodies arose on both lines, as we saw. Another important feature that arose independently on both sides was *endothermy*.

This is, roughly speaking, warm-bloodedness, using internal processes to maintain a stable body temperature that is usually warmer than one's surroundings. Warm-bloodedness is expensive, requiring large amounts of energy, but it has considerable advantages. You can stay alive and active in a broader range of environments, and muscles gain power and stamina. All those storm-like processes essential to life that buffeted us back in Chapter 2 run differently at different temperatures, and they tend to run best at temperatures higher than the ones usually handed to us by environmental conditions. With endothermy, a body and brain become a higher-energy system, consuming more oxygen.

Endothermy in its full-fledged form evolved independently in mammals and birds. But a range of related traits is seen here, not just one. In those mammals and birds, a roughly constant temperature is held by continual regulation of metabolic processes. Various other animals can keep themselves a *little* warmer than their surroundings, or do this for just part of their bodies. Behaviors such as shivering, panting, and seeking out warm or cool areas can also be combined with regulation of the internal fires. Some of the earliest steps toward endothermy may have been taken by insects, well before mammals and dinosaurs, and insects now use a range of tricks to regulate their temperature. In bees and flies, the rapid beating of wings warms the middle part of the body, and some of that heat is pushed up to the head and brain. Some meticulous work from Simon Laughlin's laboratory at Cambridge has shown that when the eyes of a fly are warmer, they can respond to motion in a more acute, fine-grained way. Events that would be seen blurrily in the cold resolve into sharpness when the fly is warm.

In the sea, warm-bloodedness is rare. It is seen today in one group of ray-finned fishes, including tuna and swordfish, and in two groups of sharks, including the great white shark. Those ray-finned fish seem to have evolved different kinds of temperature regulation, and did so several times. The tuna warm their whole bodies; swordfish only warm the eyes and brain. The effects in swordfish are similar to those seen in flies: the eyes resolve motion more finely when they are warm. The temperature of activity does matter for the cognitive, information-processing side of the the mind, and it should also then matter for experience, through its effects both on cell-to-cell interactions and on the more elusive global dynamic properties of the brain.

Long ago, some predatory marine reptiles—ichthyosaurs and others—might also have been endothermic. But most fish and all marine invertebrates (including the octopus) cannot maintain elevated body temperatures. Temperature control is easier on land than in the water, as water conducts heat away from the body more readily than air does. Staying warm is a task in which the continuities between our watery bodies and a watery medium pose difficulties, rather than helping; there are gains from being a liquid system living in air. Temperature itself on land is usually more variable, and that can be a problem, but body warmth is easier to maintain.

Body temperature in dinosaurs is a hotly debated topic. Birds are warm-blooded, but that does not imply that many earlier dinosaurs were. Birds evolved a particularly frantic, high-energy form of life. Some researchers have argued that the active life of some of the classic, bone-crunching dinosaurs would also have required warm-bloodedness, but there seems to be little consensus on how far back, and how broadly, dinosaur endothermy might run. And as we saw, warm-bloodedness is not a yes-or-no matter.

If we ask—what was dinosaur experience like? What was it like to be a dinosaur?—then birds are our best model. Birds *are* dinosaurs, after all, and the extinct dinosaurs of the Mesozoic are now seen as far more bird-like than they were a few decades ago.

Experience in a classic extinct dinosaur might be akin to experience in a large and lower-powered bird.

That link to birds brings us closer to familiar cases, but even as we approach our most local parts of the evolutionary tree, there are surprises. One of these takes us back to a theme of Chapter 6: the integration of nervous systems, especially between the two sides of the brain.

Vertebrates are bilaterian animals, as we saw earlier, following an ancient design featuring symmetry across left and right sides. Many parts of the animal are then paired between left and right, and this includes much of the brain. The two sides of the brain do not always share information as much as we might expect.

Now that we're looking at vertebrates, here is a diagram with a final fragment of the "tree of life," to orient us once again.

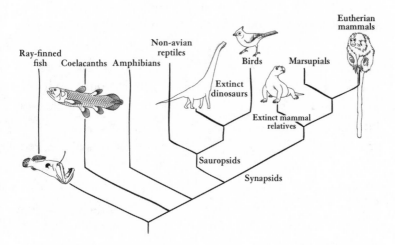

Coelacanths are some of our nearer living fish relatives. To their left are many other fish, with a deep-sea anglerfish as ambassador. Out of view beyond the anglerfish are starfish, octopuses, crabs, and so on. To the right of coelacanths are amphibians such as frogs, and then the two big branches that featured in the previous section, Synapsids and Sauropsids.

All these vertebrates inherited a brain design that initially evolved in fish. That brain is centralized in the animal's head, but in many of these animals, there is limited connection between the left and right sides. The two sides of these brains also seem to use somewhat different "styles" of processing, and have different specializations. (I mentioned this briefly in Chapter 6.) In a wide range of animals, the left side of the brain specializes in identifying food, while the right has an aptitude for social relations and threats. Cautiously, researchers sometimes suggest more general differences: the left can seem better at categorizing objects, while the right handles relationships. For example, Giorgio Vallortigara and Luca Tommasi used temporary eye patches on young chicks to get them to handle a problem using either the left side of the brain or the right (or with both sides, when they wore no patch). Chicks wearing no patch initially learned a location to find food that could be distinguished by both a landmark *and* the location's general position in the space. Then they were put into a situation with one eye or the other covered, and where the landmark and the general spatial cues no longer agreed, as the landmark had been moved. The chicks using their left eye—hence the right side of the brain—ignored the landmark and followed the general location; chicks using their right eye, hence left brain, did the opposite, going to the new location of the landmark.

Chicks without a patch also ignored the landmark, surprisingly. The right brain seemed to dominate in that case. Though the left might have good information, it can't get a word in, unless the right is taken out of the picture.

Toads show some quite odd left-right behaviors, even though they have eyes on the front of their heads, not the side, and hence have a good binocular field of vision. If prey is brought from out of view into the left side of a toad's field of vision, where most information goes to the right side of the brain, the toad usually won't strike at the prey until it has crossed into the other half of the visual field, engaging the left side of the brain. The left, again, is the side that specializes in identifying food. The opposite applies, roughly speaking, when what is seen is a competitor toad rather than food.

The specialization of halves of the brain—their particular interest in food or competition or something else—does not seem too surprising. What does seem striking is the combination of this specialization with limited connections between the left and right sides; indeed, researchers on lizards and fish, in particular, have found themselves explicitly comparing these animals to "split-brain" cases in humans. While the left and right brains of these animals have some connections, the links are rather slim: they do not have anything like the large highway-like connection, the *corpus callosum*, that is found in us. This arrangement, where the sides are both specialized and disconnected, seems to have such clear costs—food on the left will be neglected, threats from the right ignored. But those costs are apparently worth paying, because specialization brings such gains in efficiency of processing: you might be a great deal better at identifying food, as long as it's visible on the right.

Vision is the best-studied sense in this area, as is usual, but perhaps you remember the blind cave fish from Chapter 7 who navigate with their lateral lines? They make their way through complex environments by sensing the location of obstacles with their "long-distance touch." As I said above, the left side of the vertebrate brain seems to have a particular interest in *What?* questions; a number of animals use their right eye (remember the crossed wires) to inspect new objects. The blind cave fish have the same preference. When confronted with a new landmark in a navigation task, they present their right side, with its lateral lines. Once the fish are familiar with that landmark, the preference recedes. Fish have lateralized lateral lines.

I'll also leave vertebrates for a moment and mention a speculation that came from my octopus watching. If an octopus at Octopolis has another bearing down upon it in a threatening way and decides to flee, it tends to produce a pale, mottled pattern as the flight begins. But sometimes just half the animal goes pale, and the other half does not. One possibility is that the whole animal deliberately produces a half-body pattern for some reason. Another possibility is that the pattern is controlled by just one side. Perhaps one eye sees the incoming threat, and that side of the brain activates a pattern on the same side. (Octopuses do not have left-to-right crossed wires in the brain, as we do.) If so, octopuses have once again displayed on their bodies some hidden neural facts.

In vertebrates, fish, amphibians, and reptiles have the weakest connection between their left and right sides, and birds are a mixed case, still somewhat separated. Experiments have shown in some cases that if a bird learns a task—a choice of some kind—using only one eye, it can't do the task if it is made to use only the other. Then, with mammals, we reach the evolution of the corpus callosum, the large left-right connector that is severed in split-brain surgeries. I say "mammals," but even here it's not universal. I was startled to learn that no corpus callosum is present in marsupials such as kangaroos, or monotremes such as the platypus. In these cases, it appears that other connections between the two sides are playing a significant role; those Australian mammals seem to be a holdover from an earlier design. "Eutherian" mammals (the ones that used to be called "placental") are the only animals with a corpus callosum.

Working my way through the tree, it was surprising to me that less integrated designs are a normal part of life in so many animals. Earlier I said that some researchers compare the extreme examples, fish and lizards especially, to the split-brain condition in humans. In that case, as we saw in the earlier chapter, many researchers think that the people in question end up with two minds housed within one body. Might a fish or lizard, also, have two minds?

The situation here might give some support to a view of the kind I cautiously adopted in Chapter 6. A single mind might have its parts connected, in some cases, by unusual means, including pathways that loop through the external environment and back, and depend on the body's movements. These vertebrates do act as wholes; they are coherent agents, getting things done as animals need to. They perceive and act as selves. Remember all those behavioral findings from earlier chapters—remember, for example, the archerfish who can learn to hit flying insects by squirting jets of water at them, and who pick this ability up by watching other fish do it. That is quite a skill—a skill shown by the entire fish—and a skill that is cleverly acquired. The idea that hidden in the bodies of countless individual animals of this kind is a pair of minds, not one, seems to be flying in the face of the integrated way they proceed through the world.

On the other hand, some of the facts here are pretty odd, by any interpretation. For a fish, the right side of the world seems to be populated by categorized objects, while the left side has less in the way of clear objects but richer networks of relations, and is more socially "loaded." Giorgio Vallortigara, whose work with chicks I mentioned earlier, and who has done decades of work on these left-right differences, does think of this as a difference in how each side sees the world.

These left-right questions also connect to the themes of Chapter 7, which brought the large-scale dynamic features of brains into the picture. In my first split-brain discussion in Chapter 6, we had not yet contemplated the brain's rhythms and fields, but there were hints of their presence. Split-brain operations are done to reduce the severity of epileptic seizures, to stop them from spreading through the brain. A seizure is a kind of large-scale dynamic process. As the typical success of the split-brain operation indicates, separating the sides leads to less integration in the large-scale dynamic activities of a brain. This fact extends beyond seizures and the like; the slow brain waves seen in sleep are less synchronized across left and right in a split-brain patient.

Animals with more limited left-right connections in their brains will have large-scale dynamic patterns that are less unified. This sort of disconnection does not make those patterns go away, but they will be different. Those differences are probably highly relevant to experience. A split brain might in some ways work as a whole, tied together by indirect paths that draw on behaviors and bodily unity, but bonding of this kind doesn't bond everything. Even when that animal is behaving coherently as a whole, responding to what it sees and acting, experience will be different in a less connected brain. The extent of the differences will depend on questions about the respective roles of the schematic, self-related properties discussed in Chapter 5 and the dynamic properties of brains discussed in Chapter 7. It's natural to wonder, also, whether special circumstances in the life of a fish or reptile with a semi-split brain could induce a division into two minds, as seems to happen in humans in experimental settings. Perhaps the activities present in the two halves, in these non-human cases, are so different from what goes on in the much larger human halves that there's never enough to yield a two-subjects situation. In these animals, the separated upper parts of the brain are also a smaller fraction of the whole brain than the cortex is in a human.

I am not going to push much further with those questions here, saving many of them for another book that will consider, among other things, the factors that conjoin to make humans such unusual animals—language, technology, and social life, together with our large, energetic, and highly connected brains. To finish this section, I'll look at another evolutionary path that links land and sea.

Dolphins are mammals who returned to the sea—they are "toothed whales," along with sperm whales and some others. The line leading to dolphins and other whales diverged from the line leading to primates around 90 million years ago, back in dinosaur times. Dolphin ancestors seem to have returned to the sea about 49 million years ago, and the land-dwelling dolphin ancestors are relatives of the hippopotamus. The other group of whales, baleen whales, split from toothed whales about 34 million years ago.

Some dolphins create, with great skill, bubble rings in the water (a bit like smoke rings) and play with them. Videos of this behavior, taken by Diana Reiss, are remarkable but also rather poignant. The rings are perfect and the dolphins do a fair bit with them—swimming near and appreciating them. But watching this behavior produces that sense, as with fish, that if only their bodies allowed it, they could do so much more. Dolphins, like birds, gave up a good deal of manipulation for their special form of mobility. If a dolphin had hands, what could it do?

Dolphins have extremely large brains, both absolutely and in relation to body size. These brains expanded after, not before, their return to the sea. Surprisingly, there has not been a great deal of work on why this happened—much less than with primates. Dolphin brain size seems to have increased in two bursts, one early and one late. The first might be related to the evolution of echolocation—sensing by producing bursts of sound and listening to the echoes. But in general, dolphins seem to be a good case for a "social intelligence" hypothesis. Their social lives are very complicated, full of intricate alliances.

As eutherian mammals, dolphins have a corpus callosum connecting the brain hemispheres, but the dolphin's version of this connector is smaller than would be expected, given the size of the dolphin's brain. Dolphins can sleep one half-brain at a time (as noted in Chapter 6, they give themselves a natural Wada procedure). Dolphins appear to be a case where, on returning to the sea, a tendency toward integration in the mammalian brain has abated or changed course, even as those brains became very large. I wonder whether this is because of the great importance of sleep; the corpus callosum is slimmed down to allow the large-scale dynamic patterns to go their own way on each side, enabling sleep in the difficult circumstance of breathing air in water.

Wild dolphins sometimes have a strikingly close engagement with people. A few years ago I saw a lone dolphin who regularly visits Cabbage Tree Bay, the marine reserve near Sydney that was the site of much of my earlier book *Other Minds*. This dolphin, well-known along those shores, is a female—I will stay with "is," though she has not been seen there for a couple of years now, as far as I know. She lives on her own, after losing her pod some time ago, and does not seem too concerned about her unusual life. That day in the water, a lot of swimmers hung around, many just to see the dolphin. We kept a distance, but she would zoom in. She took a particular liking to a young man with red hair. When he dived down, she would come rocketing in and bring her face very close to his, repeatedly, so close that it looked like she was going to kiss him. I don't know why she singled him out for this attention. Some people seem to have a way of moving in the water—perhaps a calmness, but it seems more distinctive than this—that appeals to particular animals. I've seen this with Matt Lawrence, who discovered the Octopolis site. Something about him makes octopuses want to play, and they often end up crawling all over him. It was similar with the red-haired man and the lone dolphin.

The Roles of Land and Sea

Entering the sea, dropping below the surface, some changes are immediate—new colors, the press of water. But there is an upper layer, of variable depth depending on conditions, where life can still directly feel the energy of the sun.

Photosynthetic plankton, seaweeds of various kinds, and corals with algae as partners use this shallow sunny region. Below that layer, the sun is still usable for seeing, but provides little energy, and resources must come from living matter. A dive under high cliffs at Jervis Bay sent us through the transition sharply, the light gloomy below a thin first layer. Beside us were rock walls hung with the stationary bodies of animals filtering particles from the water. We were not that far down, but well into the next realm.

Far below any diver is the abyssal deep sea. Another octopus watcher, Bret Grasse, spent years watching dark video screens of data from remote submersibles, 2,500 feet down. The usual sight in the screens was falling snow, pale gray particles drifting into darkness. The snow was organic material from upper waters—discarded casings of plankton, other castoff parts and waste, and the remains of bodies.

Vampire squid, the animals Bret was watching for, hang suspended in a zone of very low oxygen. Few other animals can survive there at all—the vampires find safety in inhospitability. They feed by sending out long thread-like feeding tentacles, collecting particles from the snow. The screens have few visitors; Bret saw seven or eight vampire squid in four years.

The land has a third of Earth's area but about 85 percent of its species, at least among multicellular organisms. (The situation with bacteria and the like is less clear.) The crossover—the point in time after which the land dominated in diversity—occurred just in the last 100 million years, relatively recently in the timescales of this book, but the imbalance has been marked since then.

An impression that naturally arises is that although life had its early stages in the sea, once the land was colonized evolution took off in a new way. A series of papers on this theme have been written by Geerat Vermeij, a distinguished biologist at the University of California, Davis. Vermeij has worked especially on molluscs. He is unusual in his thinking, as he welcomes some controversial comparisons between evolution and human affairs: he sees evolution in terms of economic competition, arms races, and invasions. I, an amateur when writing about these enormous biological timescales, have occasionally hesitated at descriptions of these events in terms drawn from human conflict, especially in these chapters of invasion and colonization. Vermeij, in contrast, pushes these comparisons hard, not because he thinks this is harmless vivid talk, but because he thinks the analogies are close and the same principles are at work.

There is another unusual fact about Vermeij: he has been blind since the age of three. All his science, all his fine exploration of the shells of molluscs, has been done without the aid of sight.

Applying his economic mode of thinking, Vermeij argues that the land is a site of more creative evolution than the sea; it has produced more "high-performance innovations," and this might be expected. One reason is the sheer productivity of the land, with its high rates of energy flow. Another is the scope for animal action: "Activity is less constrained in air than in the denser, more viscous medium of water."

In a 2017 paper, Vermeij argues for this by going through a list of evolutionary innovations and seeing where they arose: sea, land, or both. He starts the comparison from the Ordovician, the time when animals arrived on land, and looks forward from there. His story is one in which life moves through its infancy in the friendlier zone of the seas, but evolution takes off once organisms conquer the challenges of life on land.

Some of the contrasts between Vermeij's picture and mine are a matter of perspective. He begins at a stage where the basics are in place and can be taken for granted, whereas my book is about the basics. But a few of his comparisons of land and sea are, I think, not entirely "fair" to the peculiarities of life in the sea. For example, one of the evolutionary innovations Vermeij discusses is flight—aerial locomotion. Flight, he says, was invented in both places, but earlier on land. However, all sorts of other marine animals were already flying, in a sense. On land, flight (along with some burrowing and climbing) is the only way to escape the surface and move in a fully three-dimensional mode. In the sea, swimming and drifting are inherently 3-D. Even before fish, jellyfish were flying.

Some of Vermeij's other cases, such as warm-bloodedness, seem fairer as land-sea comparisons. But I resist the conclusion that conditions in the sea "constrain innovation" (as he puts it) more than conditions on land. An alternative picture sees both as sites of creativity, and places the innovations into a historical order.

A number of huge, and necessarily early, innovations occurred in the sea: the evolution of animals and animal bodies, senses, limbs, nervous systems, and brains. The sea is the natural context for these stages. Once the animal way of living is in place, moving onto land is a possibility, with an embrace of its hothouse flux. Then further rounds of innovation are needed. Results include the mammalian and avian bodies, tight control of body temperature, new kinds of social organization, and new abilities to engineer the environment. We have the marine stages to thank for the nerves and brains through which these words are buzzing, for animal bodies, and for experience itself. But a move to land opens new doors.

10

PUT TOGETHER BY DEGREES

When I awoke at midnight, not knowing where I was, I could not be sure at first who I was; I had only the most rudimentary sense of existence, such as may lurk and flicker in the depths of an animal's consciousness; I was more destitute of human qualities than the cave-dweller; but then the memory, not yet of the place in which I was, but of various other places where I had lived, and might now very possibly be, would come like a rope let down from heaven to draw me up out of the abyss of not-being, from which I could never have escaped by myself: in a flash I would traverse and surmount centuries of civilisation, and out of a half-visualised succession of oil-lamps, followed by shirts with turned-down collars, would put together by degrees the component parts of my ego.

—MARCEL PROUST, *In Search of Lost Time*, Volume 1

1993

I remember the first time I heard whales sing underwater. They were humpback whales on Australia's Great Barrier Reef near the Whitsunday Islands, over twenty-five years ago now. I remember the face of the person I was diving with when the sound started. She turned and her eyes seemed to get as big as saucers.

The sound seemed to come from far away. It was well and truly audible, occasionally even loud, but it seemed to be carrying from some great distance. The song was quick-time, moving busily over frequencies. I think it went on for just about all of the dive, though I am not sure about that.

That part of the reef is currently a crumbling remnant of its former beauty, but that is not the point of the story. The point of the story is the memory. This is a particular kind of memory. *Episodic* memory is memory of an experience, where the memory is not just a noting of what happened—"I was at the Whitsundays and heard whales"—but something that is itself semi-experienced, or at least has an experience-like side, accompanied by visual, sonic, or other sensory images. There is something it's like to remember particular important events. I can recall a bit of the feel and rhythm of that day, though it was a long time ago. Episodic memory is an important part of human experience, and also a path to a theme that is bigger.

Elsewhere

Memory is a basic aspect of mind and cognition. One conception of memory is that it is like storage: bits of information are put away for later use. Memory has figured in earlier chapters of this book mostly in relation to learning; learning inevitably requires some sort of retention, the making of a mark in the nervous system.

Psychology distinguishes between four or five main kinds of memory. Semantic memory is memory of facts—Paris is in France. Procedural memory retains the ability to do things—how to ride a bicycle. Episodic memory is memory of experienced events. Those forms of memory can persist for the long term. There is also "working" memory, the momentary retention of ideas and images as you manipulate them.

Episodic memory is usually taken to have two main features: it is memory of particular events rather than generalities, and memory that is experienced or relived. A large and sometimes disturbing scientific literature describes how unreliable human memory is in many circumstances, and episodic memories can seem especially "cooked up."

The Canadian psychologist Endel Tulving, who named episodic memory and charted these distinctions in the 1970s and 1980s, observed that a patient with severe amnesia affecting his episodic memory, Kent Cochrane, also had another problem; he could not imagine events in the future. Cochrane was the first of several patients studied who had these paired problems. Clive Wearing, an English expert on early music, acquired amnesia in 1985 due to an infection, and ended up with largely intact semantic and procedural memories but highly impaired episodic memory, along with a distressing near-permanent sense that he had just woken up. Imagined futures, again, are as inaccessible to Wearing as remembered pasts.

Around 2007 a series of papers appeared that presented new data on these relationships and also put them together into a theory, or a cluster of theories. The new data made use of brain imaging, and found that brain regions involved in episodic memory are also active when imagining the future. The theory proposed, in different forms, that a single capacity for "mental time travel" is involved in both looking forward and looking back. In retrospect, other hints of a relationship between these things were closer at hand, and just needed to be noticed. We can, without deliberate effort, "remember" events from visual perspectives that we never experienced. You can see yourself from the outside in an episodic memory.

The new thinking about episodic memory from this time also included ideas about what purpose this kind of memory serves. These included ideas that put memory itself into the background, from a functional point of view. They suggested that the important and useful thing we do with mental time travel is simulate possible situations in order to aid planning. These situations need never have happened—they are mere possibilities, things that might come about tomorrow. Episodic memory, which looks backward, is a byproduct of this forward-looking ability. (Versions of these views are known as the *future-first* or *constructive episodic simulation* hypothesis.)

Why should we believe this? One reason is that episodic memory is so unreliable. If its role was to be a mere record, we might expect something more accurate. The combination of unreliability and vividness seen in episodic memory is part of what suggests that it is a byproduct of a skill in exploring possible futures. A forward-looking faculty brings with it an ability to cook up pasts as well.

Some statements of this view perhaps went too far in contrasting the forward-looking task of planning with the backward-looking activity of remembering. Semantic memory, too, will only be useful if it helps us make choices about future actions, and that is entirely compatible with it being a record of facts learned in the past. Semantic memory is also often found to be inaccurate, though its main role is to retain traces of the past to inform tomorrow's actions; perhaps episodic memory is not so different. Consider an episodic memory of a complicated recent event, like the bustle of a party. You can go back to this memory and—perhaps with errors—try to mine it for things you may not have noted explicitly at the time (those two people *did* spend a lot of time talking to each other).

Retention of facts need not be the sole function of either kind of memory. Another role for both kinds of memory, including their fallible side, is shaping a narrative that helps us to maintain a character, a picture of oneself. This possibility was suggested by earlier studies of the unreliability of episodic memory. That still seems reasonable, and partially distinct from the exploration of futures in planning.

A different way of interpreting many of these findings is that episodic memory, imagination, and some other abilities we have are all part of a general capacity for "offline processing," a capacity that can be directed forward, backward, or sideways (into alternative presents). In the phrase "offline processing," the "line" in question is the usual flow that runs from sensing to action. We are "off" that line when we construct and entertain possibilities without their being a direct response to what we see, and without our acting, at least right away, on where they take us. Donna Rose Addis, one of the leading psychologists in this area, has moved to a view of this kind.

This package of offline abilities, the ability to take ourselves away from what is actually around us, is an important part of the human mind. In a way, its arrival is the advent of the mind in an additional sense—the mind as free-wheeling and creative, not bound down to the here and now. Offline modeling is useful; this is a tool with practical importance in working out what to do. It also gives human experience much of its feel.

The idea that offline processing is a single but multifaceted skill has been extended to the case of dreams. After centuries of understandable treatment in religious and spiritual terms, the first detailed theory of dreams with a good basis in neuro-biology (I do not include Freud) was developed by the Harvard psychiatrists Allan Hobson and Robert McCarley in the 1970s and 1980s. They suggested that dreams result when a burst of activity comes from the brain stem, a lower and ancient part of the brain, and the cortex tries to make some sense of it. Francis Crick and Graeme Mitchison later hypothesized that dreams are junk, in a sense familiar to everyone with a computer. Dreaming is getting rid of useless and disorganized information, drag-ging it to the trash and preventing it from clogging up your brain. Against this background, some recent theories of dreams are the most sensible-seeming of all, with respect to what dreams are said to do. These views treat dreams as a kind of modeling, a turning-over and recombining of possibilities, along with replay of snippets of past experience to consolidate memory. These views treat dreaming as continuous with other offline activity, including daydreaming, exploration while dozing, and episodic memory itself.

Clearly dreams yield a rather unruly form of this explora-tion and scenario-modeling. Perhaps dreaming is a mix, with Hobson and McCarley's idea—chaotic activity getting from the lower brain to the cortex, which tries to make sense of it—in the picture as well. That activity may even be complementary to the more goal-directed side of things, as an injection of useful noise; certainly it would stir things up.

All this contributes to a picture in which human cognition includes a family of experienced or semi-experienced offline processes—internal events that feel like something, with uncertain boundaries and likely links to dreaming. What is the relation between these phenomena and the mind-body story I have been telling?

Much of the account of experience given in earlier chapters has been concerned with the embedding of a self within the world, handling it in "real time"—a here-and-now engagement with things. In an evocative phrase lurking around philosophy since the time of Martin Heidegger, and used by Andy Clark to title an influential book, much of the story so far has been about *being there*. Now we encounter the removal of felt experience from that role. First was *being there*; now there is being *elsewhere*.

This side of experience is probably not confined to humans; many other animals may have it. Dreams are a clue. Sleep itself is extremely widespread in animals and probably very old. Its function is rather unclear, but goes beyond mere physical rest. Cuttlefish, which are octopus relatives but often even more colorful, have been the subjects of two remarkable studies of sleep. The first uncovered evidence for a state in these animals that looks very similar to REM (rapid eye movement) sleep in us. In humans, quieter "slow wave" sleep alternates with more active REM periods, and REM sleep is more associated with dreaming. The second cuttlefish study found a similar alternation between sleep modes in cuttlefish, despite their vast evolutionary separation from us. In their REM-like state, cuttlefish twitch their arms, move their eyes, and produce unusual patterns on their skin. In cuttlefish, like octopuses, skin colors are controlled by the brain and the entire pattern can be changed in less than a second; skin patterns are a direct reflection of the animal's moment-to-moment brain activity. Whether or not those brain processes are experienced, as human dreams are, these creatures' skin is truly a window in.

Cephalopods, especially cuttlefish and octopuses, produce other rambling color changes and patterns with no obvious role. They do this when sitting quietly but apparently awake, as well as in sleep-like rest. Once again, the history of these animals makes all these facts even more notable. Cephalopods last shared a common ancestor with us something like 600 million years ago. Even views that propose that the common ancestor back then was more complex than has usually been assumed (views discussed in Chapter 6) hold that cephalopods built their complex brains almost from scratch, after a period of less exciting limpet-like life. The fact that these animals evolved a recognizable alternation between two modes of sleep similar to ours, not to mention the fact that they display their brain processes during the REM-like state in color on their skins, is remarkable.

Do these processes in cuttlefish do for them anything like what offline processes do for us? We have no idea. But in the case of rats, some fascinating work suggests that similar projects are pursued. "Place cells" are part of a well-studied system in the rat brain that maps the physical environments a rat has experienced. These cells are, as the name suggests, cells that fire when the rat is in a particular place. By watching the sequential activation of these cells, researchers can also trace a pathway through space that a rat is representing to itself when it is not actually moving. Rats some time ago were shown to engage in replay, during sleep, of paths they had taken earlier. That is not all they do; as well as replay, rats engage in *pre*-play, exploration of paths they have never actually taken, but leading to places where they have encountered food. They can trace out, in their brains, new paths to a goal, and the activation of one place cell after another shows that they are doing this. On waking, rats often move along the pre-rehearsed path.

For someone interested in hidden offline processing, the existence of the rat place cell system is about as miraculous as the cuttlefish's dermal windows into the soul. Do the rats also experience their replays or preplays, in the way we experience dreams? Or is this just an internal process that silently solves a problem, like those cases (at least a dozen during the writing of this book) when you wake up in the morning with a solution to last night's intractable problem present in your head, and no idea how it got there? That question about experience is very hard to answer—it is hard enough to know which nocturnal processes in our own brains are experienced, given how readily dreams are forgotten. And much of this work on offline navigation in rats has studied slow-wave sleep, rather than REM sleep, which rats do have (much less surprisingly than cuttlefish). But there is at least one good clue. The replay of paths can be compared within REM and slow-wave sleep. In slow-wave sleep, the path represented was traversed by the rat in a flash—about twenty times faster than it would take to actually walk that path. During REM sleep, the replay of the journey occurred at a more natural speed. Periods of behavioral activity that were minutes long—movement through an environment that extended for minutes—were replayed in the sleeping brain, place by place, at the same sort of speed.

This is not conclusive, but does establish a bridge between animal experience and the usefulness of offline processes, as there is in us.

Taken all together, this suggests that some other animals do have offline processing of possibilities, and offline experience, too. A lot of animals spend a considerable amount of time just sitting. When they do, I don't think it is all blank in there, and I doubt that they experience a fixed monotonous slide of the here-and-now scene, either. The self-propelled dynamic patterns in animal brains have more going on in them than that. I think many animals spend a fair amount of time elsewhere. A likely difference between human and non-human cases is not the existence of elsewhere-experience, but the extent of its deliberate control. A feature of human cognition that really does seem to differ greatly from what goes on in other animals is something psychologists call "executive control," the ability to direct oneself on a task, suppress momentary urges, and marshal one's various abilities in pursuit of a consciously represented goal. Through this side of human cognition, in concert with tools such as language as an organizer of thought, we can deliberately induce and control our offline journeys, rather than just having them happen. We can set out deliberately to some particular elsewhere, though we might also meander, drift off, and dream.

By Degrees

As we draw close to the end of the book, let's look at the overall picture coming into view. The guiding idea has been that the story of mind and felt experience comes out of the story of animal life. Animal evolution produced a new kind of being, one with new kinds of engagement with the world through the senses and action. It produced subjectivity and agency. It also produced animals who handle their dealings with the world in a way that includes a tacit sense of self. I don't think that the sense-of-self element is *the* solution to the problem, a turning on of the experiential lights. But it is important as a feature of the "shape" of sensing and action in animals, and clearly relevant to subjectivity. Back in Chapter 5, I wondered about the roles of these parts of the story—just *being* a self of the new kind, a locus of sensing and action, and having that sense of self. Part of the situation might now be put like this: Evolution, in animals, puts a premium on coherent action, and from a certain point onward, the way to act effectively *as* a self is to have a sense *of* oneself, as a unit of that kind. This sense of self is entirely tacit or implicit at first, but can become less tacit as the complexity of behavior continues to evolve.

Alongside those ideas is another. As it turned animal bodies into these centers of sensing and action, evolution installed unique features in the nervous systems that control those bodies. The basis for experience is not just a collection of cells that talk to each other in a network, but an organ with further kinds of activity and integration, through distributed rhythms and fields and perhaps other large-scale dynamic properties.

Questions remain about how these ideas fit together—about what sort of weight goes onto the sheer formation of a self engaged with the world and acting on it, what role the emerging sense of self-and-other has, and how much is due to the unique ways that nervous systems organize nature's energies. Each does its own kind of partial bridging of the gap between mind and matter, between experience and biology, but I take seriously the message of the Grothendieck passage (the lost epigraph) discussed in Chapter 1—the idea that the problem will be slowly transformed as knowledge is built around it.

This view of things is tied to nervous systems (it is "neurocentric," in a sense), and hence skeptical about the existence of non-animal minds. Why couldn't something else do what nervous systems do? Perhaps it could, but look at *what* they do, on all the many scales. There is cell-on-cell influence in a network; there is slower oscillation of a kind that synchronizes activity; there are fields spreading in space and affecting neurons in turn. Thinking about brains and experience, one can fall into one image or another. Either a brain is all switches and signals between cells (*and somehow that must be enough*), or, when its global dynamic properties come onto the scene, one can suddenly visualize a conscious cloud, a cloud of feeling extruded by the brain. In a sense, the first view makes the problem seem too hard, while the second makes it seem too easy. What is special about a brain is the combination, with all the local cell-to-cell influences—organizing stimuli from sensory surfaces, creating coordinated action—and the large-scale activity patterns as well. All this was shaped by evolution; it is not something that nature just spontaneously produces. It evolved to enable action and its control.

If materialism is true, then the storms within cells, the threading together of countless cells' activity, the perturbed rhythms of their electrical breathing, and their large-scale coordination *are* the stuff of mind. This is what we are asked to identify with—not to think that our minds are a *consequence* of this, but that our minds *are* such activities. Part of what is involved here is a kind of imaginative leap, a leap into monism—the philosophical view described back in Chapter 1 that asserts a unity in the world's basic composition, despite the apparent differences between mind and matter. The imaginative leap is to the idea that we are not *extra*, not additions to the physical world, but aspects of its workings. We are *of* these activities rather than merely tied to them or made by them.

So far, that is a view of which sorts of things have subjective experience, and why they do. But next we can ask: What *is* experience? First and fundamentally, it is the activity described above as felt *from the inside*. It is the way things feel for a system that has the right kind of activity in it. Experience is what it's like to *be* that system—not to look at it, or describe it, but to be it. It's what it's like for a pattern of activity of the right kind to be *you*.

In a bit more detail, the situation looks like this. If you are a living thing of the right kind (and awake, and so on), you have an *experiential profile*, at each moment. This is the way things feel as a whole, at that moment. That profile is affected by a huge amount of what is going on in you, in larger or smaller ways. The experiential profile changes as the moments pass.

That profile will be a very different thing in different animals. In a typical human case, though, it will include a sensory tableau—with things seen, heard, and felt—along with many other elements—mood, energy level, a tacit sense of one's body. A profile of this kind includes big differences between foreground and background. If you are attending to something—to a book, to a sound, or what is happening outside—that is foregrounded and much else recedes. This can shift quickly. You can attend instead to the hardness of the seat or the breeze from the fan, either because something has changed or just because you choose to. These can then recede again. When they recede, they are still part of your experiential profile. They are still having some effect on how you feel, as is your energy level, mood, and so on. All these can be in the background, but still contribute.

I'll now introduce another part of my picture, the last piece that

I want to put in place. This is *gradualism*, the idea that in evolution, minds and experience come slowly into view, rather than appearing in a single move.

An investigation like this one begins, inevitably, with conscious experience in its rich and complex human manifestations. That soon leads to questions about simpler cases, and where it all begins. Charting the biological features responsible for experience, we then find that many of them (perhaps all) seem to be matters of degree. Among present-day organisms, whether a nervous system is present at all seems to be a clear-cut matter, interestingly. At least in the standard view of what a nervous system is, either you have one or you don't. But nervous systems probably evolved gradually, and just about every other feature that matters to the story has borderline cases around us now.

Some changes in evolution can be quite rapid (due, for example, to mutations that affect the relative timing of stages in an organism's life), but important traits don't pop into existence as wholes; they creep into being, making their way from borderline states. In earlier chapters we often encountered cases where an animal, or some other living thing, has just a part or a sliver of the features that make for experience in us: cnidarians, gastropods, protozoa, plants. All this suggests that the origin of experience itself was a gradual matter.

Though gradual transitions look natural on the biological side, it has seemed to others that consciousness, or felt experience, cannot be like that. The lights in an animal (ant, jellyfish, squid) are either on or not: yes or no. I've been struck in recent discussions by how many philosophers want to insist on this; people say there is no possible way to see the existence of experience itself as a graded, more-versus-less matter. Initially, they say, the situation might look a bit like that—a good model might not be a simple light switch, but a dimmer switch. But even with a dimmer switch, the light is still either on a little or not on at all. Similarly, I am told, differences might exist in the complexity and richness of experience once the internal lights are on, but being on to some degree is still being on.

This picture is one to resist. If the features I've been describing are the basis for experience, and these shade off into partial cases, then subjective experience does too. Subjective experience is not something that comes into being at some stage and from that stage is just *there*; rather, its thereness is a matter of degree. It can be partly there, partly not.

Why does this seem hard to accept? First, it is easy to be trapped by some of the language we use. That idea of the "lights being on" is entirely a metaphor. People often resist giving up this image when they talk about consciousness, or they set it aside and then ease back into it. But this really is a metaphor, and in some ways a bad one, as it suggests the erroneous project of conjuring colors from the brain's darkness that I criticized back in earlier chapters. The language of "something it's like . . ." also causes problems: either there's something it's like to be you or there's nothing. Something versus nothing is a distinction that is sharp. But although I think Nagel's well-known phrasing is not a bad way of embarking on the project, gesturing toward what has to be understood, we should not be tied to this language either. It certainly doesn't show anything about how the solution has to go. And at least some of the other language that offers partial bridges between mental and physical does allow degrees and gradations. A point of view can come into being, can become definite, in a very gradual way.

In giving an evolutionary account, it can be tempting to try for a view in which consciousness arises by the crossing of a threshold. Animals slowly acquire more and more of the biological features that matter and then suddenly, in a "phase transition," the lights go on. Relatively abrupt transitions do occur on the physical side of things, and those might have a place in the story. But what can't happen is that change on the physical-biological side is smooth and gradual, and consciousness suddenly appears on top. To say that is to go back toward a view in which brain activity somehow *causes* consciousness. Instead, once we have worked out the kinds of brain activity that matter here, we can say that having that pattern of activity *is* having conscious experience.

A gradualist view gets some encouragement from the familiar phenomenon of waking from sleep. Waking is a process in which, at least sometimes, a truly gradual coming-into-consciousness seems to occur. Philosophers I've raised this example with say that although your experience when waking might be vague and weak, this must become a *fully real* vague and weak experience, from some definite point onward. I suppose they would say the same thing about a human first acquiring consciousness when they are a baby. That case is hard to access after the fact, while waking is an everyday and familiar event that is quite supportive of a gradualist view. To say this is not to claim that the evolution of consciousness over millennia is similar in other ways to a waking-up event, but waking gives us a model of borderline cases.

New language waits to be developed here, but the picture I endorse is something like this. In humans, there just *is* felt experience. But as we move toward cases with less and less of what is present in us, subjectivity grades or fades away. In some other animals, "Is it conscious?" or "Does it have experience?" does not have a yes-or-no answer; instead, the goings-on inside an animal can be *more or less experiential.*

I am referring to this view as "gradualist," but there is no single scale. If you have simpler evaluation and more complex sensing, for example, you are not more conscious overall than something with simpler sensing and more complex evaluation. The right view will probably recognize several dimensions, with differences in degree on each.

Whether or not all this seems a reasonable way of pulling together the ideas that have arisen in this book, it goes against the grain of a lot of recent work in philosophy and psychology. Here I have in mind more than the resistance to gradualism discussed just above, but a bigger difference in approach.

In much current work, people try to explain conscious experience with a detailed information-processing model, one that looks for a particular crucial stage or pathway in the brain's processing: when information goes from *here* to *there*, it is experienced, and not otherwise. Our brain is doing masses of information processing, and some of it gets onto a special path or circuit, or is given a specific kind of encoding, and then it is experienced consciously. Perhaps experience arises when information is routed to "working memory," or when it is brought into a central "workspace" or "world model" within the brain. And only very little of the information being handled by the brain at any time is handled in the consciousness-producing way.

This approach is justified in part by experiments, often fascinating ones, that seem to show there is a great deal of selectivity in conscious experience. Much of what human brains do is not conscious, and the unconscious side appears to include a lot of the basic activity needed in getting by—perception, guidance of action, even basic kinds of learning. We seem to do a lot of this "in the dark," as far as experience goes. Experiments show that we can interpret words that are presented so briefly we are not aware we've heard the word at all. We don't know we've heard it, but the word's meaning can affect how we respond to other things. Similarly, people with some kinds of brain injury can deal visually with things around them while claiming they are entirely unaware of those objects—this is called "blindsight."

I respect those experimental findings, but I disagree with how they are often interpreted and put into an overall picture. It is pretty common in this approach, for example, to say that your conscious experience can only contain one thing at a time. The French neuroscientist Stanislas Dehaene, whose lab did a number of the experiments described above, believes this, for example. You can only be conscious of one thing, though you can switch from one to another very quickly. That is in some ways an extreme position in this family of views—others think more than one thing can be present, but it all has to be brought together along that crucial path or routed to that crucial place in the brain, whatever it might be.

When I think about the role of factors like mood, energy level, and so on, which I emphasized above, these "narrow pathway" views of experience seem so surprising that I suspect they are talking about something entirely different from what I was talking about, in my own sketch of experience. Perhaps there are two things here. One is the overall experiential profile—everything about what it's like to be you, right now. The second is what you are conscious *of* at the moment—whatever is in the foreground, being attended to. To me, the word "conscious" belongs more with the second than the first, in part because the word suggests that there's some object involved: Are you conscious . . . ? OK, what are you conscious *of*? The idea of an "experiential profile" is supposed to be different from this. A lot of what makes a difference to experience—a lot of what contributes to what it feels like to be you right now—is nowhere near that level of focus.

If an experiential profile is a broader, more expansive thing in this way, how much of what is going on inside a brain contributes to it? The answer initially suggested by my view is: *everything*, even if much of what is going on makes a contribution that is tiny. This is because any alternative seems to head back toward dualism, toward the idea that experience is the product of some particular part of a brain's activity, something made for us, offered up for us, by our brains. Instead, our experience *is* that activity, so everything about that activity must matter to some extent. However, the situation is not quite so simple. Though everything—every detail—of the pattern of activity that is your mind will matter to experience, not everything going on within the confines of your body, or your skull, has to be a part of that pattern. The resulting picture is still very far from narrow pathway views of subjective experience, though. Many things going on inside a person will make a tiny difference to pacing, attention, or mood, for example, and those things *are* part of experience.

The narrow pathway views I am using as a contrast to my own also tend to push away from the idea of widespread animal experience. This is partly because they are shaped, understandably, by findings about the human case, and the theories that people come up with are theories of how particular parts of our own brains do what has to be done. In addition, these views suggest a picture that I think influences people on this matter, even when the picture is not explicit. The experiments I mentioned above seem to show that there is a longish list of basic things that a human can do unconsciously—various kinds of perception, and processing of what we sense. If so, it seems that we might put all these things into a package, and produce an animal that can sense the world and respond to it fairly well while being entirely unconscious. After all, this animal is only doing things that we know can be done unconsciously. It would then feel like nothing to be that animal. Next, this conclusion might be applied to various actual animals, whose abilities are similar to the hypothetical one. This would not be a good argument, however. The experiments the argument is based on were all done with conscious human beings, even though the people were not conscious of everything going on inside them. From these experiments we are learning that, when you are a conscious human being, a surprising amount can be done deep in the background. That does not show that everything could be in the background at once. For any normal and wakeful human being, there is *something* it's like to be that person, even if much is being done behind the scenes. The same may apply to a great many animals as they make their way through the world.

I said above that many of the leading ideas about consciousness and experience in recent work are shaped by findings about our own case. That is understandable, as these views usually have a basis in experimental work on humans, who can follow instructions and report on what they see and feel. Work of this kind should indeed tell us a lot about how experience works in us. But what about the rest? In some cases, a setup that roughly matches the relevant human brain parts is found in other animals (birds, fish), and this opens the possibility of extending the view. But we also have to deal with animals further from us, including many of the animals in this book. The mistake that can then be made is to say: *this* (as studied in humans) is what experience *is*, and if you don't have this going on, you can't have experience. The alternative is that if your brain is different from ours, your experience is different, but not absent. Once we accept that there is probably some form of experience in hermit crabs, octopuses, and so on, there's a need for a broad view—a broad account of what they have, and we have, that makes us all experiencing beings. That is what I have tried to make progress on here. In addition, I think that the broad approach taken here furnishes clues about human experience, our own case. It reminds us of those primordial, indistinct elements of experience that accompany the focused, attentive awareness most obvious in psychological experiments. Human experience is a mixture of old and new.

Consequences

Suppose that these ideas are on the right track—incomplete, but right as far as they go. What follows, for other questions presently debated? What are some consequences? I will look at two.

I said in an earlier chapter that many people look at bees and flies, for example, as little flying robots, and not subjects of experience. Some readers might have wondered: Why am I so down on the robots? Why write them off? Perhaps present-day robots cannot have experiences, but future ones might.

Related questions concern artificial intelligence (AI) systems of many kinds, systems where a mind is supposed to exist that is not housed within a body in the usual manner, but "realized" in a pattern of interactions in a computer program. I have in mind here attempts at what is called "strong" AI; these are computer programs that are built not merely to behave or solve problems as someone with a mind would, but programs that are meant, when they are running, to *be* a mind. Further, if a mind can exist in a pattern of software interactions, we might expect one day to upload some of them, perhaps including our own minds, to a computing "cloud." Physical computers will be needed in this scenario, but a mind might move from computer to computer, as information does in cloud computing systems today. Our thoughts and experiences, even if existing only in localized bodies now, might flit from machine to machine after the upload.

If the ideas in this book are on the right track, you cannot create a mind by programming some interactions into a computer, even if they are very complicated and modeled on things our brains do. The view of this book is opposed, to varying degrees, to the thinking behind many of these AI projects.

These AI projects are motivated by the idea that minds exist in patterns of interaction and activity. Usually, those patterns are found in our brains, but the same patterns, it is said, could exist in other physical devices. The view I'm defending here does, in a way, agree that minds exist in patterns of activity, but those patterns are a lot less "portable" than people often suppose; they are tied to a particular kind of physical and biological basis.

One objection that is often made to strong AI is that in these computer programs, you might be able to *represent* a pattern of interactions of the kind seen in a brain, but that is not the same as having those interactions be *present* in the computer. They are merely encoded, written down, and that is not enough. This is an important objection, and AI advocates often dismiss it too glibly. But there are some kinds of activity, related to what our brains do, that might be given actual existence in a computer without too much difficulty. Suppose that a brain was just a signaling and switching network, where neuron A triggers the firing of neurons B and C, neuron C affects D, E, and F, and so on, and this is all that happens. Then it may be that as long as something in a computer plays the role of A (affecting B and C), something else plays the role of B, and so on, you have all you need—the brain's pattern of activity can be present, not merely represented, in the machine. But neurons—and brains—do more than this. The objection that AI programs merely represent what brains do, and do not *do* what brains do, is much more acute for the large-scale dynamic features of brains. These, too, would have to actually be present in the computer. It would not be enough to work out some equations that describe what these rhythms and waves (and so on) do in the brain, and run those equations in the machine. The machine has to actually *have* these patterns present within it.

If the goal is *a* mind rather than a human mind, then these patterns would not need to be exactly like they are in our brains; they might be merely similar. But give some thought to what would be required to get anything like this to happen in a machine. Think of the activities that give rise to the rhythms and fields in the brain. These are tiny ebbs and flows of ions (charged particles) over membranes, combining to yield coordinated oscillations in particular parts of the brain. Even if we set aside the brain's fields, it may be very difficult indeed to have a system with anything like the brain's dynamic patterns that is not otherwise physically similar to a brain.

This is another good moment to think through the meaning of the kind of materialism I am describing in this book. At the outset, you find yourself full of conscious thoughts. How could all that just be a brain? Look at it, a gray blob. That can't be enough. Someone says: No, no; it's not just that object, it's the activities inside the brain, and you can't see those. You ask: What activities? The person replies: A lot of signaling and switching, very intricate. You are supposed to say: OK, and you are supposed to feel that it now makes sense. But is the situation really much better? Then you see the whole; you see, along with these cell-to-cell influences, the rhythms and fields, the patterns of electrical activity being modulated by the senses. Then, speaking for myself at least, the situation seems entirely different. Those activities *can* be me, my thoughts and experiences, my reliving past events and imagining the future. It's no longer hard to credit.

Present-day computers contain a kind of logical sliver, a tiny fraction, of what goes on inside us. Computers are often designed to create illusions of agency and subjectivity, and they do that well. If we start from a device containing some fast and reliable logical processers attached to a large store of memory, housed in a stable unit that is fed all the power it needs, this remains a completely different object from a brain, and from a living organism, no matter how it is programmed. Perhaps, in the future, artificial systems could be built from different materials and could have more brain-like operations. The result might be a kind of artificial life, or at least something closer to it than current AI systems. The problem here is not artificiality, the fact that AI systems are made by us rather than evolution. The problem is the need to have the right kinds of things going on inside.

In the area of AI, then, the biggest disagreement I have is with "upload" scenarios. The idea that an upload of a computer program could have experiences just like yours, and could be a continuation of *you*, is just a fantasy. You are a very different thing from any pattern of activity that might wander from one computer to another in the cloud. Again, future machines might be different from present ones, and artificial life may one day come. But given the technologies on the table now, there could not be a process that takes your experiences from their biological basis in a living body, and has them continue—as more of *you*—in the cloud.

Upload scenarios are the most unlikely. At the other end of the spectrum are scenarios that imagine future robots with genuinely brain-like control systems inside them. That might, one day, yield not only artificial intelligence but artificial experience, too.

If sentience exists in roughly the way I say it does, how might this change our behavior toward animals and other living things? The question is too big to tackle in detail here, but one thing this book does is give us reason to extend some *consideration* to many more animals than usually receive it now. The welfare of many more animals becomes part of the picture, something to take into account. Extension of consideration is not the same as extension of rights, or trying to establish some kind of equality of status; the aim of this book is not to install mosquitos, midges, and aphids as fellow citizens, or to argue that we need to radically change our behavior toward animals of those kinds. In a way, extending consideration is not a very big step in itself, but it is *a* step, and one that is warranted.

One part of my view leads to an awkward place. Once we give up on the traditional divide between humans and all other animals, when thinking about welfare and ethical questions, there is a very reasonable temptation to look for a new divide, a new border. We might say: sentient beings are on one side and everything else is on the other. Some recent proposals have said that once there is a reasonable chance that an animal is sentient, we should err on the side of protecting its interests. Much can be said in favor of this attitude, and in some contexts it can easily be applied. But if a gradualist view is true, then there may be a lot of cases, including various invertebrates, where the question "Is it sentient?" does not have a definite answer even in principle. We will have to find new ways to think about those cases. The question of very different varieties of experience is also important here. I had suspected for a time that various insects have experience of a sensory kind, but no evaluative experiences, such as pain and stress. If that is so, it would have a big effect on questions about their welfare. But I now think, as we saw in Chapter 8, that this view of insect experience may well be wrong.

A more specific issue is also raised. All through this book I have used information that was gained, directly or indirectly, from experiments that were cruel in various ways. The science is often reported very simply, especially once the findings have been processed by review articles and the like: here is how the animal works; here is what is inside it. But behind the scenes is often a lot of suffering. How should we think about that very work, the work that got us here? Some of it I have no problem with. Earlier I said that about Elwood's work with crabs. In other situations in my reading, I encounter an experiment and just turn away, especially work on monkeys and cats. This book has not included a lot of material about mammals, so it has not had to confront those hardest cases too much. The offline rats in this chapter were getting us close. That information could hardly be more interesting—it is almost a miracle that they have this system inside them, and that it can be studied. Questions about how the brain works are being transformed by that research, and it has resulted in Nobel Prizes. I am very glad that I know these findings, but it's all been harder for the rats. One might hope that as more is learned, what we learn induces us to back off rather than press forward.

In these discussions of animal experiments, we are often

talking about very small numbers. Do a handful of animals in experiments much matter, in this context? The questions we should ask do not always depend on numbers; there is also the question of what relationship we want to have to other sentient animals. Sadism, for example, is bad even when the numbers involved are tiny. I do not think that experiments of the sort discussed here are usually sadistic, in a literal sense of that term—taking pleasure in another's pain. But once it is pretty clear that something results in pain for an animal, if you keep doing it just because it is interesting to learn about what is going on . . . I am not sure how to describe this, but it's a bad relationship to bring into being. Often it is reasonable to hope that knowledge gained will lead to a reduction of suffering elsewhere, and that is important. In this area I am most haunted by the fact that practical progress often comes as an unanticipated consequence of basic research done for other reasons. This fact must be confronted, and it makes the accounting very hard. Even when reasoning of this kind dominates a situation, much can be done to keep experiments informative while reducing the probable experiential effects on the animals. One might be glad that knowledge was gained while not wanting, now that we know, work of some kinds to ever be done again.

The Shape of Mind

Last year, finishing a dive in the bay that has been the scene of many events in this book, I swam into the shallows and paused when two small scraps of seaweed began fighting vigorously in front of me, with tufts flying. On closer inspection, they were tiny decorator crabs, using algae rather than sponge as clothes.

They seemed quite irritated with each other, and in both cases uninterested in backing down. Neither gave ground while I watched. They stayed close and occasionally flailed seaweedily at each other. This was an initially surprising, though ultimately comprehensible, display of agency.

What is the overall layout of mind in the world? Where and when is it found? How much of it lies around us? Two starkly different pictures can be compared; we might think of these as desert and jungle. The desert view holds that mind exists hardly anywhere. The world is mentally barren over almost its whole extent, including the living world. The view of Descartes, sketched in the first chapter, is a position of this kind, but one does not have to be a dualist to have a view like this; a lot of people have had similarly sparse pictures. Perhaps humans and some other mammals have minds, but once we leave this group, mind is left behind. The rest of the world is mentally blank. Even most of its living parts are mindless shells.

The opposite vision is a jungle, with mind everywhere or nearly everywhere. The most extreme version of this view is panpsychism, with soul-like powers seen even in atoms. A near-jungle is also seen in a view of the kind that says all living things—plants, bacteria—have feelings. Then sentience is nearly everywhere on Earth, excepting only completely lifeless zones.

The truth is between these, neither desert nor jungle. Its shape can be explored by moving once again over the genealogical tree of life on Earth, the tree produced by evolution, now over 3 billion years old. First, and surprisingly, some activities that are mind-like in a broad sense are all over the tree, perhaps on every branch and stem. Sensing and responding are everywhere. No living cells are entirely oblivious to what is going on around them. Though we don't know about life's earliest stages, what is now called "minimal cognition" appears to exist over much or all of the tree's extent.

This is surely a surprise. The view of Aristotle seems closer to how we might have expected things to turn out. Aristotle thought that all life has a "nutritive" soul, a drive to keep itself alive, and only animals also have a "sensitive soul," the ability to perceive and respond. Humans, in addition, have a rational soul. It seems that things could have turned out like that—with much of life sustaining itself blindly, and the senses creaking open for the first time in animals. Instead, a sensitive soul is as widespread and perhaps as early as any.

From this starting point, different evolutionary lines travel their own paths. On one branch, the animal body arises, a new means for action, and on this branch, nervous systems tie bodies together in new ways. Within this branch, on some of its sub-branches and twigs, animals start to move rapidly, manipulate objects, and perceive the objects they act on. This branch generates the animal way of being.

What shape does sentience, *felt* experience, consciousness in the broadest sense of the term, have in this picture? Once sensing and minimal cognition are recognized everywhere, and once animals like octopuses and crabs are seen as sentient, an unnerving vista can open before you. Before you stretches a gradual slope that leads into plants, fungi, non-neural animals, protists, and bacteria. If sentience has to come into being gradually, why isn't *this* the road on which it appears? Why doesn't minimal cognition imply minimal sentience? If subjectivity is an important idea in making sense of the evolution of the mind, doesn't everything with minimal cognition have a kind of subjectivity, a way things *seem* to it, and so on?

That question has become the biggest recurring uncertainty as I have worked on this book. It opens a path not to panpsychism, but to what can be called *biopsychism*, the idea that all life is sentient. (The term was invented by Haeckel, who wrestled with this question, though I am modifying his meaning a little.) However, I think this is a mistake. Minimal cognition is, after all, present in bacteria, and when you look at what they do and how they do it, feeling just seems to not be part of the picture. Going back and forth over these questions, I've come to a sense of the importance of nervous systems, and—again—the unique way they organize nature's energies.

This does not resolve everything, as we still have almost all animals before us. From here, one option has it that although precursors to experience exist in much of the animal part of the tree, felt experience arose several times in different evolutionary lines. This happened in arthropods, some molluscs, and vertebrates—perhaps more than once in some of those groups. Perhaps it also happened in additional groups I've not discussed here, but a lot of other invertebrates (corals, bryozoans) just do not have what it takes. Sentience arose in animals from its absence, in this view, and did so at least a handful of times.

A second possibility is that a primordial form of experience arose just once, longer ago, early in animal evolution. That form was in place for the evolutionary radiations we have charted in this book, and developed in its own way down various different lines.

Which is it? The attempt to resolve this question encounters every possible problem; it relies on events lost in the past, involves working out what is going on inside animals with very different nervous systems, and also working out what the simplest forms of experience might *be*, what it means to say that sentience was on the scene at some very early stage.

We've burrowed just about to a point where I can go no further. If I had to bet, I'd bet on the first of those options above, but I am not sure what the hunch is based on, and maybe the truth is some blend of the two, a blend that can't yet be expressed with the language we have on hand. Still, I'd bet on a view in which definite sentience exists not just in vertebrates, but in some groups very distant from our own: at least in cephalopods and some arthropods. From there, looking backward, we find no sudden switching-on of the lights along those evolutionary lines, but a gradual process by which the *self* becomes more definite, internal goings-on become more experiential, and subjectivity takes shape. The only organisms that feel their lives have nervous systems, therefore only animals. As you move along a reef, or through a forest on land, sensing and minimal cognition exist in all the life around you. Some of that life is organized into selves who are subjects of experience as well. Sentience is not absolutely everywhere, even within life. But there is a lot of it, from sea angels (perhaps) to sea dragons (for sure). The world is *fuller*, more replete with experience, than many people have countenanced.

In working through this summary and its uncertainties, I have again been writing of the *being-there* kind of experience, taking in your life as it happens. Another side of experience, as we saw earlier in this chapter, is the ability to leave the here and now, and find oneself elsewhere. This is a liberatory gift from evolution, and one given not only to us. Many other animals probably have some of this kind of experience, especially in dreams. But in humans, this capacity is brought under more deliberate control.

This elaboration, the growth of offline processing, makes the mind into something more like a *thing*. What does that mean—a thing as opposed to what? I mean that the story told in earlier chapters is one that might best be told in ways that avoid reifying (thing-ifying) the mind itself, treating the mind as an object. An animal is a physical object, as is its brain, but its mind is less an object than an aspect of its workings and the activities within. Once offline, matters are a little different. Now a mind becomes something like an *arena*, one where we rehearse options, make things up, and manipulate sonic and visual images. The philosopher John Dewey, back in the 1920s, wrote that the mind has both a practical role, guiding action, and along with this, a role as an "esthetic field." It is a place where scenes can be built, tales can be told, and present circumstances set aside. I assume that the most basic forms of experience are not like this, and *being there* comes first. But when this other dimension appears, it is the advent of the mind in a further sense.

Thirty years ago, as a student, I went to a conference and attended a group of talks about the Austrian philosopher Ludwig Wittgenstein. Wittgenstein, working in the early and mid-twentieth century, is the person who argued in most detail, and with the most influence, that the picture of the mind that philosophers have usually worked with is founded on illusions. In a phrase made famous by Gilbert Ryle, who was much influenced by Wittgenstein, philosophers have thought of the mind as a "ghost in the machine"—a spectral controller, mysteriously housed in a material body. Once you have this picture of the mind, both Wittgenstein and Ryle thought, you have no chance of making sense of things. The picture generates endless and spurious problems—problems we don't need to solve, but simply abandon.

Following this theme, one speaker at the meeting (Crispin Wright) talked about the philosophical error of treating the human mind as something like a "walled garden," a secret place that one person—the mind's owner—can observe privately, and with no possibility of error. For Wittgenstein, minds could not be anything like this. Minds, instead, are present in *what people do*, in their behavior.

The "walled garden" metaphor was supposed to indicate an error, a wrong turn. But as the idea was turned over in discussion, a couple of speakers, even at this event, seemed to sense its appeal. And ordinary people, not just philosophers, often talk in a way that suggests a view like this—they talk of the mind as a private realm. That was a point to pause over, because Wittgenstein had wanted to respect everyday talk before it has been contaminated by philosophy. He thought that philosophers are much more confused in these matters than other people (though it does not take much for a person to start thinking like a philosopher).

People might talk about minds as hidden realms, but such a view can't be right, the speakers thought. Things just can't be that way. I think, instead, that things *are* that way, or something like that way, in us. This is not a general feature of sentience, something that would be present in every animal that experiences its life. But in our case, that is how it is. We create an arena, a field, filled with both things that arise and things we furnish. We enter a garden.

NOTES

1. PROTOZOA

5 *It was preserved in alcohol and sent to the biologist T. H. Huxley*: My main source for the *Bathybius* affair is Philip F. Rehbock, "Huxley, Haeckel, and the Oceanographers: The Case of *Bathybius haeckelii*," *Isis* 66, no. 4 (1975): 504–33. Huxley's 1868 paper is called "On Some Organisms Living at Great Depths in the North Atlantic Ocean," *Quarterly Journal of Microscopical Science* (n.s.) 8 (1868): 203–12. Here Huxley says he thinks the substance is "protoplasm" and christens it *Bathybius Haeckelii* (using a capital "H," as I have it in the text, rather than the usual lower-case for the species name).

6 *Haeckel was delighted with both the discovery and christening*: For Haeckel, I draw on the biography by Robert J. Richards, *The Tragic Sense of Life: Ernst Haeckel and the Struggle Over Evolutionary Thought* (Chicago: University of Chicago Press, 2008), and a good recent sketch, Georgy S. Levit and Uwe Hossfeld's "Ernst Haeckel in the History of Biology," *Current Biology* 29, no. 24 (2019): R1276–84. Back-to-back with that one, on pp. R1272–76, is a paper about Haeckel's famous illustrations: Florian Maderspacher's "The Enthusiastic Observer—Haeckel as Artist," including controversies over their accuracy.

Haeckel, like many biologists of that time, believed in a racial hierarchy that had white Europeans at the top. He has sometimes been linked to the development of Nazism in Germany. Richards debunks these claims (without trying to claim that Haeckel's views were entirely enlightened) in "Ernst Haeckel's Alleged Anti-Semitism and Contributions to Nazi Biology," *Biological Theory* 2 (2007): 97–103. Haeckel's top rank of humans, for example, included Jews and Berbers (*Berber* and *Juden* are alongside *Romanen* and *Germanen*). Richards also notes that Haeckel became a close friend of the pioneering homosexual activist and sex researcher Magnus Hirschfeld, who dedicated his book *Natural Laws of Love* (1912) to Haeckel.

6 *Both were also eager to press on to questions*: The most famous passage is a cautious thought in an 1871 letter to J. D. Hooker: "It is often said that all the conditions for the first production of a living organism are now present, which could ever have been present.—But if (& oh what a big if) we could conceive in some warm little pond with all sorts of ammonia & phosphoric salts,—light, heat, electricity &c present, that a protein compound was chemically formed, ready to undergo still more complex changes, at the present day such matter w^d be instantly devoured, or absorbed, which would not have been the case before living creatures were formed.—"

(Darwin to J. D. Hooker, Down, Kent, February 1, 1871, Darwin Correspondence Project, darwinproject.ac.uk/letter/DCP-LETT-7471.xml).

6 *Haeckel was convinced that the spontaneous generation of life*: According to Rehbock, Huxley denied that his work supported that view.

6 *When the Swedish botanist Carl Linnaeus devised a new scheme of classification*: Linnaeus's *Systema Naturae* was published in many editions from 1735 onward. The later editions included animals as well as plants, and then minerals.

7 *In 1860, the British naturalist John Hogg argued*: See Hogg, "On the Distinctions of a Plant and an Animal and on a Fourth Kingdom of Nature," *Edinburgh New Philosophical Journal* (n.s.) 12 (July–Oct. 1860): 216–25. For Hogg, as I say in the text, the boundaries between the living realms were vague while the boundary between living and non-living was sharp; in his diagram, he drew that line especially firmly.

7 *Hogg's term,* Protoctista, *was later shortened by Haeckel to the more modern* Protista: Even that latter term is now seen as questionable, as it does not pick out a single definite branch of the tree of life (Protista is a "paraphyletic" grouping). A lot of terminology used in this book is controversial for the same reason. But it is not easy to write about these topics without using terms like "fish" and "crustacean," which raise the same issues.

7 *In the framework of Aristotle, developed over two millennia earlier*: See especially Aristotle's *De Anima*. The interpretation of this work is controversial; I am treating Aristotle's framework as non-dualist, but there are more dualist readings of Aristotle, and *De Anima* contains plenty of puzzles. See Christopher Shields, "The First Functionalist," in *Historical Foundations of Cognitive Science*, ed. J-C. Smith (Dordrecht, The Netherlands: Kluwer, 1990), 19–33.

A comment from Justin Smith: "Before the early modern period, to deny souls to animals would have been to deal in paradox. The word 'animal,' after all, is simply an adjectivized form of the Latin noun 'anima,' which is to say 'soul.'" Justin E. H. Smith, "Machines, Souls, and Vital Principles," in *The Oxford Handbook of Philosophy in Early Modern Europe*, ed. Desmond M. Clarke and Catherine Wilson (Oxford, UK: Oxford University Press, 2011), 96–115.

8 *For René Descartes, an especially influential figure*: I draw on Gary Hatfield's "René Descartes," in *The Stanford Encyclopedia of Philosophy*, ed. Edward Zalta, Summer 2018, plato.stanford.edu/archives/sum2018/entries/descartes. Here, again, there are interpretive controversies, and Descartes did not publish all of his thoughts on the topic. Hatfield: "In mechanizing the concept of living thing, Descartes did not deny the distinction between living and nonliving, but he did redraw the line between ensouled and unensouled beings. In his view, among earthly beings only humans have souls. He thus equated soul with mind: souls account for intellection and volition, including conscious sensory experiences, conscious experience of images, and consciously experienced memories." Thanks also to Alison Simmons for help with these issues.

In the text I contrast Aristotle's view with Descartes's. An important stage between them is the "scholastic" framework, which merged Aristotle with Christianity, a merge with consequences for views of the soul. Thomas Aquinas was a central figure in this period. The *Stanford Encyclopedia of Philosophy* article on Aquinas by Ralph McInerny and John O'Callaghan is, again, helpful (plato.stanford.edu /entries/aquinas).

NOTES

9 *They called it "protoplasm"*: Here I draw extensively on Trevor Pearce's paper "'Protoplasm Feels': The Role of Physiology in Charles Sanders Peirce's Evolutionary Metaphysics," *HOPOS: The Journal of the International Society for the History of Philosophy of Science* 8, no. 1 (2018): 28–61. The paper is nominally about the philosopher C. S. Peirce, but it covers much more. The quote from William Carpenter is taken from Pearce's paper.

Huxley's claim that "organization is the result of life, not life the result of organization" is quoted in Rehbock's paper about *Bathybius*, and is taken from Huxley's *Hunterian Lectures on the Invertebrata* (1868), as reported by the *British Medical Journal*. According to Pearce, Haeckel was initially cautious about these questions about the mind, but from the mid-1870s he began to attribute a kind of sentience to matter itself: "[E]very atom has sensation and the power of movement," quoted in Pearce.

9 *When people looked inside cells, it seemed that not enough organization was present*: A long-standing philosophical tradition had invited us to think of ordinary matter as containing hidden worlds of intricate forms, perhaps infinitely divisible. The seventeenth-century philosopher Gottfried Leibniz argued that matter must be composed in that way. Leibniz had looked through one of van Leeuwenhoek's microscopes when he visited Holland, though he claimed that he had more general reasons to insist on realms within realms. The idea of hidden structure at this scale was at least on the table. But one suspects that people looking at cells through microscopes in Darwin and Huxley's time, if they were aware of these speculative pictures, perhaps didn't take them seriously. You are, after all, looking at a tiny transparent blob, and that transparent blob seems to be doing amazing things. Thus the temptation toward protoplasm.

10 *Then came the* Challenger *expedition*: Some of Haeckel's most beautiful illustrations represent specimens from this expedition; see *Art Forms from the Abyss: Ernst Haeckel's Images from the Challenger Expedition*, ed. Peter J. le B. Williams et al. (Munich: Prestel, 2015). Amy Rice has suggested that *Bathybius* might, after all, have been organic, though the remains of seasonal plankton rather than a special kind of life ("Thomas Henry Huxley and the Strange Case of *Bathybius haeckelii*: A Possible Alternative Explanation," *Archives of Natural History* 2 (1983): 169–80).

10 *Haeckel, more committed to* Bathybius *as a missing link, hung on*: See his "Bathybius and the Moners," *Popular Science Monthly* 11 (October 1877): 641–52. Here he also says almost the same thing that Huxley is quoted as saying, above: "Hence, life is not a result of organization, but *vice versa*."

11 *But over 100 million ribosomes could fit on the period*: In "How You Consist of Trillions of Tiny Machines," *The New York Review of Books*, July 9, 2015, Tim Flannery says, "As many as 400 million ribosomes could fit in a single period at the end of a sentence printed in *The New York Review*." Four hundred million? I had to try to reconstruct the number. Here it is, as best I can. Doing a simple comparison of areas (ignoring questions of overlap or wasted space), eukaryotic ribosomes are about 25 nanometers in diameter—25 millionths of a millimeter. A circle with that diameter has an area of about 500 nm². A period is about one-third of a millimeter in diameter, which makes for an area of about 85 billion nm². That is about 170 million ribosomes per period, in area. Give or take different period sizes and ribosome arrangements, we are indeed in the right place.

13 *As Thomas Nagel put it in 1974*: This is from Nagel's "What Is It Like to Be a Bat?," *The Philosophical Review* 83, no. 4 (1974): 435–50.

14 Panpsychism *holds that all matter, including the matter in objects like tables*: Nagel's defense is "Panpsychism," in his *Mortal Questions* (Cambridge, UK: Cambridge University Press, 1979), 181–95. Galen Strawson is an emphatic defender of the view; see "Realistic Monism: Why Physicalism Entails Panpsychism," *Journal of Consciousness Studies* 13, no. 10–11 (2006): 3–31. David Chalmers is sympathetic to a relative of the view that he calls "panprotopsychism"; see "Panpsychism and Panprotopsychism," in *Consciousness in the Physical World: Perspectives on Russellian Monism*, ed. Torin Alter and Yujin Nagasawa (Oxford, UK: Oxford University Press, 2015). A clear and simple exposition is an interview with Philip Goff by Gareth Cook in *Scientific American*, January 14, 2020, scientificamerican.com/article/does-consciousness-pervade-the-universe.

14 *Huxley was attracted to another unorthodox view*: The view is called *epiphenomenalism*, and his defense (not easy to interpret in some ways) is in "On the Hypothesis that Animals Are Automata, and Its History," an address given in 1874, in his *Collected Essays*, vol. 1 (Cambridge, UK: Cambridge University Press, 2011), 199–250.

15 *This sense of arbitrariness is related to something the philosopher Joseph Levine has called*: See "Materialism and Qualia: The Explanatory Gap," *Pacific Philosophical Quarterly* 64 (1983): 354–61. Huxley is sometimes credited with an early statement of the problem, but I think he's expressing something less specific: "How it is that anything so remarkable as a state of consciousness comes about as a result of irritating nervous tissue, is just as unaccountable as the appearance of the Djin when Aladdin rubbed his lamp." *Lessons in Elementary Physiology* (London: Macmillan, 1866), 193.

16 *Monism is a commitment to an underlying unity in nature*: The term has been used for a number of views, all in the same family. Haeckel called himself a "monist"; his panpsychism was a kind of monism. See "Our Monism: The Principles of a Consistent, Unitary World-View," *The Monist* 2, no. 4 (1892): 481–86.

17 *If I was not a materialist I'd be a neutral monist*: This is discussed in my "Materialism: Then and Now," forthcoming in a collection of papers about David Armstrong's theory of the mind and the development of materialism in the twentieth century.

17 *Here is a quote from the physician and essayist Oliver Sacks*: This is from "The Abyss," *The New Yorker*, September 24, 2007.

18 *Perhaps we will be pushed to a view like this in the end*: Setting aside questions about animals and considering only the case of humans, the neuroscientist Björn Merker describes some children he has studied and interacted with who have a tragic condition called *hydranencephaly*, in which the cortex and many other parts of the brain are just about wiped out, often by a stroke suffered at the fetal stage. These children are seriously disabled in a range of ways, and probably have nothing like the mental life that most people have. But is there no experience there at all? Merker thinks this is unlikely, as evidenced by smiles and laughter, by what seems to be a fleeting but real engagement these children have with familiar people. Merker thinks there is no reason to believe that because they have no cerebral cortex their experience is totally blank. Merker's arguments seem good ones to me. His paper is "Consciousness Without a Cerebral Cortex: A Challenge for Neuroscience and Medicine," *Behavioral and Brain Sciences* 30, no. 1 (2007): 63–81. Antonio Damasio has also argued that experience does not depend on the cortex; see Damasio and Gil B.

Carvalho, "The Nature of Feelings: Evolutionary and Neurobiological Origins," *Nature Reviews Neuroscience* 14 (2013): 143–52.

20 *The passage is by Alexander Grothendieck*: The passage appears in French in his *Récoltes et Semailles*, p. 553. The French version appears on this website: ncatlab.org /nlab/show/Récoltes+et+semailles. The usual reference for discussion and an English translation of this passage is Colin McLarty, "The Rising Sea: Grothendieck on Simplicity and Generality," in *Episodes in the History of Recent Algebra (1800–1950)*, ed. Jeremy J. Gray and Karen Hunger Parshall (Providence, RI: American Mathematical Society, 2007). The translation I give is a little different (with assistance from Jane Sheldon). I am not a mathematician and do not claim to follow Grothendieck's mathematical work.

21 *Given these new associations, it seemed wrong to begin the book that way*: Here is a word about the Melville passage that does begin the book. John Wycliffe was an English theologian, an early critic of the Catholic church, in the fourteenth century. He died of natural causes and was buried, but thirty years later the Pope directed that his body be dug up and burned, with the ashes thrown into a river. In the first American edition of his book, Melville had (Thomas) Cranmer in place of Wycliffe. Cranmer was another English religious reformer, from over a century later during the Reformation, who was executed by burning. Critics think that Melville himself made the change from Cranmer to "Wickliff," who appeared in the English edition, and it was a correction. The English edition also omits the word "Pantheistic" here, but some later critical editions restore it, in effect blending the American and English versions. I am grateful to John Bryant for his help with this issue.

2. THE GLASS SPONGE

25 *A sponge garden often begins just below*: A few chapter titles echo musical works that played some role accompanying the writing. This one refers to a 2003 CD by Loren Chasse and Jim Haynes, working as Coelacanth.

26 *Inside a cell, events occur on the* nanoscale: For much of the material in the next couple of pages, I draw on Peter M. Hoffmann's book *Life's Ratchet: How Molecular Machines Extract Order from Chaos* (New York: Basic Books, 2012), also on Peter B. Moore, "How Should We Think About the Ribosome?," *Annual Review of Biophysics* 41 (2012): 1–19, and Derek J. Skillings, "Mechanistic Explanation of Biological Processes," *Philosophy of Science* 82, no. 5 (2015): 1139–51.

28 *The origin of life occurred fairly early in the history of Earth*: For an accessible exploration of recent ideas, see Nick Lane's *The Vital Question: Why Is Life the Way It Is?* (London: Profile, 2015).

28 *The Taming of Charge*: This section title echoes a classic book by Ian Hacking, *The Taming of Chance*, about the history of probability theory (Cambridge, UK: Cambridge University Press, 1990). The taming of chance, in a different sense, is also part of what is going on here (as Hoffmann's *Life's Ratchet* discusses).

29 *Here is the inimitable Richard Feynman*: in *Lectures on Physics*, vol. 2, chap. 1, "Electromagnetism," feynmanlectures.caltech.edu/II_01.html. The Feynman *Lectures on Physics* is now available free (legally) online in its entirety at feynman lectures.caltech.edu/index.html.

32 *If bacteria invented transistors, what were they doing with them?*: See Peter A. V. Anderson and Robert M. Greenberg, "Phylogeny of Ion Channels: Clues to Structure

and Function," *Comparative Biochemistry and Physiology Part B* 129, no. 1 (2001): 17–28; Kalypso Charalambous and B. A. Wallace, "NaChBac: The Long Lost Sodium Channel Ancestor," *Biochemistry* 50, no. 32 (2011): 6742–52. For the transistor comparison, Fred Sigworth, "Life's Transistors," *Nature* 423 (2003): 21–22; and for signaling within biofilms, Arthur Prindle et al., "Ion Channels Enable Electrical Communication Within Bacterial Communities," *Nature* 527 (2015): 59–63.

35 *Living activity itself is a pattern that exists embedded in an energetic flow, one that begins and ends outside of the organism*: Comments by John Allen at a National Academy of Sciences' Arthur M. Sackler Colloquium in 2014 also influenced my thinking here. The nature of living systems—the way they exist in a to-and-fro of electrochemical traffic—inevitably makes them sensitive to external events.

35 *My colleague Maureen O'Malley expressed this well*: in a 2017 email.

35 *Sensing, in at least the most basic forms, is ancient and everywhere*: Pamela Lyon's papers have detailed and provocative discussions of simple forms of sensing. The "ground floor" is one-factor signal transduction systems in bacteria; here an internal controller responds to stimuli that happen to make their way in from outside, without a receptor or sensor on the boundary of the cell. See her "The Cognitive Cell: Bacterial Behavior Reconsidered," *Frontiers in Microbiology* 6 (2015): 264.

36 *The term "Metazoa" was introduced in the late nineteenth century by Ernst Haeckel*: This was in his *Anthropogenie oder Entwickelungsgeschichte des Menschen* (Leipzig: Wilhelm Engelmann, 1874).

36 *Animals are made up of many cells living as a unit*: I say this, and then I say "the word 'animal' refers to any organism found on a particular branch of the genealogical tree, regardless of how it lives." Is there a tension there? Yes, in a way. If, somewhere on the animal part of the tree, we found a single-celled organism, that would count as an animal, by the more official definition I give here. No actual animal is known to have gone backward from a multicellular to a single-celled state, but there is a case that comes close. Myxozoans are tiny parasites of fish and worms. They were initially thought to be protists (like paramecia). They are not single-celled but close to it, with just a few cells at most of their stages. They turn out to be cnidarians, relatives of corals and anemones, massively simplified. See Elizabeth U. Canning and Beth Okamura, "Biodiversity and Evolution of the Myxozoa," *Advances in Parasitology* 56 (2004): 43–131. Another point along the same lines: I said "the word 'animal' refers to any organism found on a particular branch of the genealogical tree." Fine, but which branch exactly? In modern biological classification, all branches can be named. In a sense, all branches deserve names. Why not use "animal" for a smaller branch, one that does not include (say) sponges? This would be fine, as long as everything on that smaller branch is included. Sometimes a term like "Eumetazoa" is used for a more restricted branch of that kind.

37 *The genealogical network that animals are part of—the "tree of life"—is not always tree-shaped*: It is a simplification, especially, to talk of a "tree" in the case of unicellular organisms such as bacteria. Talk of a "network" of life, that in some places is tree-shaped, would be more accurate.

38 *Well before animals arose, the cytoskeleton had initiated a new regime*: Here I was helped by discussion with Patrick Keeling, at a 2014 National Academy of Sciences' Sackler Colloquium. The evolution of the cytoskeleton made it possible for some organisms to simplify their metabolic chemistry and invest instead in an active, mobile life. This sounds like an animal trait, but these are unicellular organisms.

38 *The eukaryotic cell itself came to exist in this way*: See John Archibald, *One Plus One Equals One: Symbiosis and the Evolution of Complex Life* (Oxford, UK: Oxford University Press, 2014), for the development of ideas on this point.

39 *This possibility was also first sketched by Ernst Haeckel*: In his "Die Gastraea-Theorie, die phylogenetische Classification des Thierreichs und die Homologie der Keimblätter," *Jenaische Zeitschrift für Naturwissenschaft* 8 (1874): 1–55.

39 *In addition, our guts contain countless living bacteria*: As the chapter says, this point is not usually part of the Gastraea theory. A new paper has raised it, and it seems a potentially significant idea to me. See Zachary R. Adam et al., "The Origin of Animals as Microbial Host Volumes in Nutrient-Limited Seas." The paper is not yet in a journal: peerj.com/preprints/27173. The paper does not make a connection to Haeckel's Gastraea.

Associations of other kinds between the earliest animals and bacteria have been discussed a good deal; see Margaret McFall-Ngai et al., "Animals in a Bacterial World, a New Imperative for the Life Sciences," *Proceedings of the National Academy of Sciences USA* 110, no. 9 (2013): 3229–36, and Rosanna A. Alegado and Nicole King, "Bacterial Influences on Animal Origins," *Cold Spring Harbor Perspectives in Biology* 6, no. 11 (2014): a016162. I think this modified Gastraea idea is a little different.

41 *The animals that contain these clues are a trio: sponges, comb jellies, and placozoans*: A good source here is Casey W. Dunn, Sally P. Leys, and Steven H. D. Haddock, "The Hidden Biology of Sponges and Ctenophores," *Trends in Ecology & Evolution* 30, no. 5 (2015): 282–91. For placozoan peculiarities, Bernd Schierwater and Rob DeSalle, "Placozoa," *Current Biology* 28, no. 3 (2018): R97–98, and Frédérique Varoqueaux et al., "High Cell Diversity and Complex Peptidergic Signaling Underlie Placozoan Behavior," *Current Biology* 28, no. 21 (2018): 3495–501.e2. For a sample of ongoing debates about the shape of the tree, see Paul Simion et al., "A Large and Consistent Phylogenomic Dataset Supports Sponges as the Sister Group to All Other Animals," *Current Biology* 27, no. 7 (2017): 958–67. In case a reader wondered why, back in the first pages, I said that only the sponges on that first tour lacked nervous systems, this is because placozoa, which also lack nervous systems, would not have been visible, though they may well have been present.

42 *Historically, sponges have been seen as the most important of the living clues*: Some useful papers: Sally P. Leys and Robert W. Meech, "Physiology of Coordination in Sponges," *Canadian Journal of Zoology* 84, no. 2 (2006): 288–306; Leys, "Elements of a 'Nervous System' in Sponges," *The Journal of Experimental Biology* 218 (2015): 581–91.

44 *The Hexactinellida, or glass sponges, explore in their bodies this chapter's themes of unity and selfhood*: For a full treatment: Sally P. Leys, George O. Mackie, and Henry M. Reiswig, "The Biology of Glass Sponges," *Advances in Marine Biology* 52 (2007): 1–145. See also James C. Weaver et al., "Hierarchical Assembly of the Siliceous Skeletal Lattice of the Hexactinellid Sponge *Euplectella aspergillum*," *Journal of Structural Biology* 158, no. 1 (2007): 93–106; this paper has some great images.

45 *The drawings of these tiny parts below, by Rebecca Gelernter*: The originals were done by the German zoologist Franz Eilhard Schulze, who was also the first to describe a placozoan. The originals appear in F. E. Schulze, *Report on the Hexactinellida Collected by H.M.S. 'Challenger' During the Years 1873–1876* (Edinburgh: Neill, 1886–87).

NOTES

46 *A wide and rather wonderful range of possibilities has been raised*: See Werner E. G. Müller et al., "Metazoan Circadian Rhythm: Toward an Understanding of a Light-Based Zeitgeber in Sponges," *Integrative and Comparative Biology* 53 (2013): 103–17: "[W]e propose that this photoreception/phototransduction process might function as a nerve-cell-like signal transmitting system"; Franz Brümmer et al., "Light Inside Sponges," *Journal of Experimental Marine Biology and Ecology* 367 (2008): 61–64: "[S]ponges have a light transmission system, and can harbour photo-synthetically active microorganisms in deeper tissue regions. . . . [S]ponge spicules in living specimens transmit light into deeper tissue regions"; Joanna Aizenberg et al., "Biological Glass Fibers: Correlation Between Optical and Structural Properties," *Proceedings of the National Academy of Sciences, USA* 101, no. 10 (2004): 3358–63: "Such a fiberoptical lamp might potentially act as an attractant for larval or juvenile stages of these organisms and symbiotic shrimp to the host sponge."

3. THE ASCENT OF SOFT CORAL

50 *When the cameras were retrieved*: Two other researchers, Dave Harasti and Steve Smith, contributed to the study design and writing. The paper is Tom R. Davis, David Harasti, and Stephen D. A. Smith, "Extension of *Dendronephthya austra-lis* Soft Corals in Tidal Current Flows," *Marine Biology* 162 (2015): 2155–59. Something like 70 percent of these corals have died off over the last year or so. The reason is unclear; Meryl Larkin is studying the decline, with reports to come.

51 *Corals are* cnidarians: Some papers I used here are Thomas C. G. Bosch et al., "Back to the Basics: Cnidarians Start to Fire," *Trends in Neurosciences* 40, no. 2 (2017): 92–105, and D. K. Jacobs et al., "Basal Metazoan Sensory Evolution," in *Key Transitions in Animal Evolution*, ed. Bernd Schierwater and Rob DeSalle (Boca Raton, FL: CRC Press, 2010).

51 *Many cnidarians have a complicated life cycle*: I had a look at cnidarian life cycles in "Complex Life Cycles and the Evolutionary Process," *Philosophy of Science* 83, no. 5 (2016): 816–27.

52 *John Lewis, a Canadian biologist, looked at thirty octocoral species*: See John B. Lewis, "Feeding Behaviour and Feeding Ecology of the Octocorallia (Coelenterata: An-thozoa)," *Journal of Zoology* 196, no. 3 (1982): 371–84. Octocorals seem to be a problematic group for historical reconstruction. See Catherine S. McFadden, Juan A. Sánchez, and Scott C. France, "Molecular Phylogenetic Insights into the Evo-lution of Octocorallia: A Review," *Integrative and Comparative Biology* 50, no. 3 (2010): 389–410.

53 *These actions may have been an important spur to the evolution of multicellularity*: See Susannah Porter, "The Rise of Predators," *Geology* 39, no. 6 (2011): 607–608, and John Tyler Bonner's work over many years. His book *First Signals: The Evolu-tion of Multicellular Development* (Princeton, NJ: Princeton University Press, 2001) influenced my thinking about all these matters. In thinking about the transition to multicellularity, it is important to picture an *active* cast of characters at the earlier stage. Single-celled organisms can move and hunt. If you can grow large, you can live untroubled by the hunters, and multicellularity is a good way to become large. Later, multicellular organisms themselves can become active and predatory, as they did in the Cambrian. The Cambrian might be akin to a larger-scale replay of an earlier world of antagonistic engagement, with the Ediacaran as a quieter time

when action was being reinvented at a new spatial scale. Partially distinct from this predation-avoidance path is a path that might lead to sponges, and in a different form, to land plants; multicellularity also makes it possible to fix oneself in a favorable spot and live as a tower, letting food come to you.

54 *What I am highlighting here, marking as a special invention, is action that involves coordination*: This is a central theme in the work of Alvaro Moreno, Argyris Arnellos, and their collaborators. See Arnellos and Moreno, "Multicellular Agency: An Organizational View," *Biology and Philosophy* 30 (2015): 333–57; and "Integrating Constitution and Interaction in the Transition from Unicellular to Multicellular Organisms," in *Multicellularity: Origins and Evolution*, ed. Karl J. Niklas and Stuart A. Newman (Cambridge, MA: MIT Press, 2016); and also Fred Keijzer and Argyris Arnellos, "The Animal Sensorimotor Organization: A Challenge for the Environmental Complexity Thesis," *Biology and Philosophy* 32 (2017): 421–41.

54 *The medusa stage is often seen as a later addition*: This is by no means certain. See Antonio C. Marques and Allen G. Collins, "Cladistic Analysis of Medusozoa and Cnidarian Evolution," *Invertebrate Biology* 123, no. 1 (2004): 23–42, and David A. Gold et al., "The Genome of the Jellyfish *Aurelia* and the Evolution of Animal Complexity," *Nature Ecology & Evolution* 3 (2019): 96–104.

55 *Nervous systems evolved early, perhaps just once, perhaps a couple of times*: This has become quite controversial, due to uncertainty over the shape of the tree. See Gáspár Jékely, Jordi Paps, and Claus Nielsen, "The Phylogenetic Position of Ctenophores and the Origin(s) of Nervous Systems," *EvoDevo* 6 (2015): 1, and Leonid L. Moroz et al., "The Ctenophore Genome and the Evolutionary Origins of Neural Systems," *Nature* 510 (2014): 109–14.

56 *When a cell excites—a sudden shift in its electrical properties—this event is usually confined to that one cell*: There are exceptions; "gap junctions" connect some cells more directly.

56 *What makes nervous systems in the full animal sense special*: Three thought-provoking papers about the origins of nervous systems are George O. Mackie, "The Elementary Nervous System Revisited," *American Zoologist* 30, no. 4 (1990): 907–20; Gáspár Jékely, "Origin and Early Evolution of Neural Circuits for the Control of Ciliary Locomotion," *Proceedings of the Royal Society B* 278 (2011): 914–22; and Fred Keijzer, Marc van Duijn, and Pamela Lyon, "What Nervous Systems Do: Early Evolution, Input-Output, and the Skin Brain Thesis," *Adaptive Behavior* 21, no. 2 (2013): 67–85. Jékely, Keijzer, and I also organized some of these issues in "An Option Space for Early Neural Evolution," *Philosophical Transactions of the Royal Society B* 370 (2015): 20150181.

57 *Another feature closely associated with the nervous system in evolutionary terms is muscle*: This is emphasized in the works by Moreno, Arnellos, and Keijzer cited above. As these papers also note, a further innovation important in bringing the animal body into being is *epithelium*. Epithelium consists of cells in the form of sheets, with cells oriented together and often signaling to their neighbors. These sheets act as boundaries and can be folded to produce complex forms. Epithelium both enables the sealing of the body and provides a material for constructing internal shapes and passageways. Our bodies are origami-like constructions fashioned from the repeated folding of these sheets of cells. In a sponge, whose body has only a partial form of epithelium, the environment is everywhere; seawater suffuses the

NOTES

body. In a cnidarian, or in us, there is a distinct interior environment: the body is set apart from the rest of the world.

58 *But cnidarians are "skinnier," so to speak, on the sensory side*: See Bosch et al., "Back to the Basics: Cnidarians Start to Fire," and Jacobs et al., "Basal Metazoan Sensory Evolution," cited above, and also Natasha Picciani et al., "Prolific Origination of Eyes in Cnidaria with Co-Option of Non-Visual Opsins," *Current Biology* 28, no. 15 (2018): 2413–19.

58 *A medusa orients itself in the water by means of organs containing small crystals, called* statocysts: This ends up making them sensitive also to sound: Marta Solé et al., "Evidence of Cnidarians Sensitivity to Sound After Exposure to Low Frequency Noise Underwater Sources," *Scientific Reports* 6 (2016): 37979.

59 *Sensing has its raison d'être in the control of action*: Why is it only this way around? Might we just as well, or also, say that action has its raison d'être in the control of sensing? No, there is an asymmetry here. Action gains food and reproductive opportunities. Action does also affect what you perceive, and a good deal of action is aimed at controlling what is sensed, but surviving and reproducing matter in a more fundamental way. Action is not *just* the control of perception.

For related reasons, I am skeptical about ambitious versions of the "predictive processing" framework, a framework that sees the fundamental role of cognition and action as reducing surprise or uncertainty—as seen in Andy Clark, *Surfing Uncertainty: Prediction, Action, and the Embodied Mind* (Oxford, UK: Oxford University Press, 2015), and also in Karl Friston's work, such as his "The Free-Energy Principle: A Unified Brain Theory?," *Nature Reviews Neuroscience* 11 (2010): 127–38. Organisms can adaptively opt to increase the uncertainty of their experience, for example by moving into a riskier environment, if it enables them to act in ways that yield sufficiently valuable payoffs.

59 *All through this part of the book, I am influenced by the thinking of the Dutch psychologist and philosopher Fred Keijzer*: Fairly early in my thinking about the early evolution of behavior, I went to a talk given by Fred—knowing nothing about him, but intrigued by his title—at a conference in Europe. Philosophers do not really look a particular way, but Fred looked more like a good tennis player than a philosopher. His talk was a challenge to my habits of thought. A mix of biology and philosophy, it presented a different way of thinking about the evolution of nervous systems, also about animal life and the relations between philosophical and scientific ideas in this area. He was interested in how philosophical pictures—not just pictures offered *by* philosophers but pictures of an implicitly philosophical kind in the minds of scientists, whether they approve of philosophy or not—shape scientific work. The work in the talk also drew on his collaborations with Pam Lyon and Marc van Duijn.

I have cited a few of Fred's papers above, but see also "Moving and Sensing Without Input and Output: Early Nervous Systems and the Origins of the Animal Sensorimotor Organization," *Biology and Philosophy* 30 (2015): 311–31.

62 *The first period of time from which we have definite fossil evidence of animals*: Many thanks again to Jim Gehling of the South Australian Museum, who has helped me often with Ediacaran issues.

62 *This was confirmed in 2018 by a student named Ilya Bobrovskiy*: The paper is Ilya Bobrovskiy et al., "Ancient Steroids Establish the Ediacaran Fossil *Dickinsonia* as One of the Earliest Animals," *Science* 361 (2018): 1246–49.

63 *Some Ediacaran organisms look a fair bit like present-day "sea pens"*: For discussion, see

292

Shuhai Xiao and Marc Laflamme, "On the Eve of Animal Radiation: Phylogeny, Ecology and Evolution of the Ediacara Biota," *Trends in Ecology and Evolution* 24, no. 1 (2009): 31–40, and Ed Landing et al., "Early Evolution of Colonial Animals (Ediacaran Evolutionary Radiation–Cambrian Evolutionary Radiation–Great Ordovician Biodiversification Interval)," *Earth-Science Reviews* 178 (2018): 105–35.

64 *A lot of Ediacarans were initially called "jellyfish," by Reg Sprigg*: Sprigg's now-famous first paper on the topic is "Early Cambrian (?) Jellyfishes from the Flinders Ranges, South Australia," *Transactions of the Royal Society of South Australia* 71 (1947): 212–24.

64 *Sex was almost certainly present*: See Emily G. Mitchell et al., "Reconstructing the Reproductive Mode of an Ediacaran Macro-Organism," *Nature* 524 (2015): 343–46.

64 *These were distinguished by a young biologist, Ben Waggoner*: The paper that introduces the names is "The Ediacaran Biotas in Space and Time," *Integrative and Comparative Biology* 43, no. 1 (2003): 104–13. My discussion of these stages and of recent work on the Ediacaran draws extensively on Mary L. Droser, Lidya G. Tarhan, and James G. Gehling, "The Rise of Animals in a Changing Environment: Global Ecological Innovation in the Late Ediacaran," *Annual Review of Earth and Planetary Sciences* 45 (2017): 593–617. The illustration on page 69 is partially modeled on one in this paper.

65 *In a fortunate etymological confluence*: Waggoner since has done work on old myths. He might have had his eye on King Arthur and his time on the isle of fruit trees.

65 *Sponges in general are puzzling*: For the candidates, see Erik A. Sperling, Kevin J. Peterson, and Marc Laflamme, "Rangeomorphs, *Thectardis* (Porifera?) and Dissolved Organic Carbon in the Ediacaran Oceans," *Geobiology* 9 (2011): 24–33, and Erica C. Clites, M. L. Droser, and J. G. Gehling, "The Advent of Hard-Part Structural Support Among the Ediacara Biota: Ediacaran Harbinger of a Cambrian Mode of Body Construction," *Geology* 40, no. 4 (2012): 307–10. Joseph P. Botting and Lucy A. Muir, in "Early Sponge Evolution: A Review and Phylogenetic Framework," *Palaeoworld* 27, no. 1 (2018): 1–29, suggest, on the other hand, that sponges may have been absent from the Ediacaran.

66 *their branching-upon-branching design has a "fractal" organization*: See Sperling, Peterson, and Laflamme, "Rangeomorphs, *Thectardis* (Porifera?) and Dissolved Organic Carbon in the Ediacaran Oceans," and Jennifer F. Hoyal Cuthill and Simon Conway Morris, "Fractal Branching Organizations of Ediacaran Rangeomorph Fronds Reveal a Lost Proterozoic Body Plan," *Proceedings of the National Academy of Sciences, USA* 111, no. 36 (2014): 13122–26.

67 *This fossil, christened with that unbelievably difficult name*: Charles D. Walcott, of Burgess Shale fame: "Dr. Fitch proposed the genus *Helminthoidichnites* for *tracks resembling those of worms*." Asa Fitch was a physician. Also a speech therapist? See Walcott's "Descriptive Notes of New Genera and Species from the Lower Cambrian or Olenellus Zone of North America," *Proceedings of the National Museum* 12, no. 763 (1889): 33–46. There are now several different fossils called *Helminthoidichnites* from different times and places, with tentatively different names; it's not thought that "Helminthoidichnites" refers to the same track-making animal in every case. For the Ediacaran burrowers, see James G. Gehling and Mary L. Droser, "Ediacaran Scavenging as a Prelude to Predation," *Emerging Topics in Life Sciences* 2, no. 2 (2018): 213–22. The new candidate for the track-maker is described in Scott D. Evans et al., "Discovery of the Oldest Bilaterian from the Ediacaran of

South Australia," *Proceedings of the National Academy of Sciences USA* 117, no. 14 (2020): 7845–50.

70 *The evidence is good that nervous systems evolved in a body with something like a radial design*: Cnidarians, which have a "radial" design, also have a simpler cell-layer organization than bilaterians, an organization that is thought to have arisen earlier. Ctenophores remain a wild card. They are often described as having a "biradial" design that is a kind of mix of the other two. And a few ctenophores are seafloor crawlers that look like flatworms. That is intriguing; perhaps they deserve a close look?

71 *I often wonder whether, with respect to the evolution of behavior and interaction between animals*: A present-day cnidarian clue discussed earlier in the chapter becomes relevant here again. Nematocytes, the stinging cells seen in today's cnidarians, are thought to have evolved early in that line, before the main groups of cnidarians (corals, anemones, jellyfish) split from each other. If that split was in the Ediacaran, as some work has argued, then at least simple forms of those stingers were around back then. These need not have been anything like the high-speed harpoons seen in the present, but their existence suggests that predation of some sort was already present. This applies whether the stingers were offensive or defensive weapons.

72 *How useful a clue are flatworms?*: For the rather intense debate about the evolution of the simple "acoel" flatworms, see Ferdinand Marlétaz, "Zoology: Worming into the Origin of Bilaterians," *Current Biology* 29, no. 12 (2019): R577–79, and Johanna Taylor Cannon et al., "Xenacoelomorpha is the sister group to Nephrozoa," *Nature* 530 (2016): 89–93. A good book about the polyclad flatworms I discuss in more detail is Leslie Newman and Lester Cannon's *Marine Flatworms: The World of Polyclads* (Clayton, Australia: CSIRO Publishing, 2003).

73–74 *But quite a few flatworms are mimics*: See Newman and Cannon's *Marine Flatworms* for the mimicry. Nudibranchs are the subject of a large literature, in part because they are so popular with divers. One place to start is David Behrens's *Nudibranch Behavior* (Jacksonville, FL: New World Publications, 2005).

4. THE ONE-ARMED SHRIMP

78 *But all these animals are arthropods*: On the evolutionary relationships, see David A. Legg, Mark D. Sutton, and Gregory D. Edgecombe, "Arthropod fossil data increase congruence of morphological and molecular phylogenies," *Nature Communications* 4 (2013): 2485. On the fossils, Edgecombe and Legg, "The Arthropod Fossil Record," in *Arthropod Biology and Evolution*, ed. Alessandro Minelli et al. (Berlin: Springer-Verlag, 2013), 393–415; this book is full of good material. For their brains, see Gregory Edgecombe, Xiaoya Ma, and Nicholas J. Strausfeld, "Unlocking the early fossil record of the arthropod central nervous system," *Philosophical Transactions of the Royal Society B* 370 (2015): 20150038.

79 *The Cambrian, beginning about 540 million years ago*: The literature is vast and here are a couple of articles I found interesting, with different perspectives: Erik A. Sperling et al., "Oxygen, ecology, and the Cambrian radiation of animals," *Proceedings of the National Academy of Sciences USA* 110, no. 33 (2013): 13446–51; Rachel Wood et al., "Integrated Records of Environmental Change and Evolution Challenge the Cambrian Explosion," *Nature Ecology & Evolution* 3 (2019): 528–38 (a no-explosion view).

80 *They seem to have just drifted offstage*: Alternative scenarios are discussed in Simon A. F. Darroch et al., "Ediacaran Extinction and Cambrian Explosion," *Trends in*

Ecology & Evolution 33, no. 9 (2018): 653–63. These authors favor a view in which there is some continuity between Nama Ediacarans and the first Cambrian animals, and the bigger shift was between the White Sea and Nama stages within the Ediacaran.

81 *The largest arthropod that has ever been found*: This is *Aegirocassis benmoulae*, who grew to at least seven feet (over two meters). Peter Van Roy, Allison Daley, and Derek Briggs, "Anomalocaridid Trunk Limb Homology Revealed by a Giant Filter-Feeder with Paired Flaps," *Nature* 522 (2015): 77–80. Strictly, it might be a near relative of arthropods (a stem arthropod).

83 *The same sort of change occurred with sensing*: See Roy E. Plotnick, Stephen Q. Dornbos, and Junyuan Chen, "Information Landscapes and Sensory Ecology of the Cambrian Radiation," *Paleobiology* 36, no. 2 (2010): 303–17, and Andrew R. Parker, "On the Origin of Optics," *Optics & Laser Technology* 43 (2011): 323–29. Todd E. Feinberg and Jon M. Mallatt, *The Ancient Origins of Consciousness: How the Brain Created Experience* (Cambridge, MA: MIT Press, 2016) also has a rich discussion of this topic. They emphasize the use of spatial structure, especially the formation of internal "maps," as a resource in complex sensing.

85 *I am quoting from a paper coauthored by Roy Caldwell*: This is S. N. Patek and R. L. Caldwell, "Extreme Impact and Cavitation Forces of a Biological Hammer: Strike Forces of the Peacock Mantis Shrimp *Odontodactylus scyllarus*," *The Journal of Experimental Biology* 208 (2005): 3655–64.

85 *These shrimp, like the tiny ones I saw in Indonesia*: Charming titles are still possible in science: "Hawksbill Turtles Visit Moustached Barbers: Cleaning Symbiosis Between *Eretmochelys imbricata* and the Shrimp *Stenopus hispidus*," by Ivan Sazima, Alice Grossman, and Cristina Sazima, *Biota Neotropica* 4, no. 1 (2004): 1–6.

86 *I don't know of research on the use of these feelers themselves*: As this book was being edited, a paper came out about the brain of this very species: Jakob Krieger et al., "Masters of Communication: The Brain of the Banded Cleaner Shrimp *Stenopus hispidus* (Olivier, 1811) with an emphasis on sensory processing areas," *Journal of Comparative Neurology* (2019): 1–27. The paper is anatomical and does not look at behavior, but it has a lot of interesting material. The antennae are covered in tiny sensors, and it seems this species would have rich chemical sensing.

86 *Animals of this general kind, though—crayfish and flies—have been shown to have*: For a review, see Trinity B. Crapse and Marc A. Sommer, "Corollary Discharge Across the Animal Kingdom," *Nature Reviews Neuroscience* 9 (2008): 587–600. As they say: "It is this coordination between the two systems [the sensory and the active] that makes it possible to analyze the world while moving within it."

In 1950 two German scientists, Erich von Holst and Horst Mittelstaedt, wrote a classic paper about this phenomenon, well before it had been found in the wide range of animals in which it is now known, and they coined some good terminology. A problem that any active animal has is distinguishing what they called *exafference* from *reafference*. *Ex*afference (with emphasis on the first syllable) is any effect on your senses that is due to an *external* event of some sort. *Re*afference is any effect on your senses that is due to your own actions. An animal might try to make this distinction by looking for some way that self-caused events look different in themselves, but the obvious way to do it is to interpret sensory information in a way that takes into account one's own actions. See von Holst and Mittelstaedt, "The Reafference Principle: Interaction Between the Central Nervous System and the Periphery," in *Behavioural Physiology of Animals and Man: The Collected Papers of*

Erich von Holst, vol. 1, trans. Robert Martin (Coral Gables, FL: University of Miami Press, 1973).

David C. Sandeman, Matthes Kenning, and Steffen Harzsch, "Adaptive Trends in Malacostracan Brain Form and Function Related to Behavior," is another interesting paper about marine arthropod behavior, in *Nervous Systems and Control of Behavior*, ed. Charles Derby and Martin Thiel (Oxford, UK: Oxford University Press, 2014). A relevant passage: "Motion is perceived by an eye when an image moves across the photoreceptive cells in the retina. This can result from the movement of an object in front of the stationary eye or by the movement of the eye past a stationary object. As long as animals remain stationary, any motion in the visual field can be confidently assumed to be extrinsic to itself. During voluntary movements however, the situation is more complex because it is then necessary to distinguish between self-induced and extrinsic image movements. A strategy to confront this problem—development of eyes that can move independently of the body—has evolved several times. This allows an image to be stabilized, within limits, onto one area of the eye." Mantis shrimp have tremendously movable eyes. Banded shrimp also have mobile eye stalks, though they are smaller and, I assume, less mobile.

87   *The neuroscientist Björn Merker sets things up this way*: See his "The Liabilities of Mobility: A Selection Pressure for the Transition to Consciousness in Animal Evolution," *Consciousness and Cognition* 14, no. 1 (2005): 89–114. Just below, where I talk of "another way to look at this situation," I am indebted to Fred Keijzer.

88   *Hearing is very different, though*: For the differences between senses, in relation to reafference, and a different view of their significance, see J. Kevin O'Regan and Alva Noë, "A Sensorimotor Account of Vision and Visual Consciousness," *Behavioral and Brain Sciences* 24, no. 5 (2001): 939–1031. Aaron Sloman's "Phenomenal and Access Consciousness and the 'Hard' Problem: A View from the Designer Stance," *International Journal of Machine Consciousness* 2, no. 1 (2010): 117–69, is interesting on this theme.

88   *The only animal reported to pass a version of it*: See Masanori Kohda et al., "If a Fish Can Pass the Mark Test, What Are the Implications for Consciousness and Self-Awareness Testing in Animals?," *PLOS Biology* 17, no. 2 (2019): e3000021. I say "a version" because of the argument that the fish do something considerably simpler here than dolphins and other mirror alumni. See Frans B. M. de Waal's commentary, "Fish, Mirrors, and a Gradualist Perspective on Self-Awareness," *PLOS Biology* 17, no. 2 (2019): e3000112. These reservations seem to be reflected in the final title of the Kohda et al. paper, which also has an editor's note about this question. The results look pretty good to me.

89   *Some especially significant work has been done by Robert Elwood*: See Mirjam Appel and Robert W. Elwood, "Motivational Trade-Offs and Potential Pain Experience in Hermit Crabs," *Applied Animal Behaviour Science* 119, no. 1–2 (2009): 120–24, and Barry Magee and R. W. Elwood, "Shock Avoidance by Discrimination Learning in the Shore Crab (*Carcinus maenas*) Is Consistent with a Key Criterion for Pain," *Journal of Experimental Biology* 216, pt. 3 (2013): 353–58. Elwood reviews this work in his "Evidence for Pain in Decapod Crustaceans," *Animal Welfare* 21, suppl. 2 (2012): 23–27. Michael Tye discusses this work (and it makes the title) in his *Tense Bees and Shell-Shocked Crabs: Are Animals Conscious?* (Oxford, UK: Oxford University Press, 2016).

For other interesting hermit crab behaviors, see Brian A. Hazlett, "The Behavioral Ecology of Hermit Crabs," *Annual Review of Ecology and Systematics* 12 (1981): 1–22. For example, hermits have long been described as "fighting" over shells, with larger individuals evicting smaller. But in at least some species, the "loser" of the fight also gains a more suitable shell; "when the 'defender' would not profit by an exchange, it only rarely vacated its shell." That suggests, Hazlett says, that the bouts of shell rapping by "aggressors" may consist of mutually advantageous communication about shells, not displays of crab size and aggression.

91 *But as Elwood replies, crustaceans also don't have brains with visual areas that are anything like ours*: See "Is It Wrong to Boil Lobsters Alive?," *The Guardian*, February 11, 2018.

92 *an early study of hermit-anemone associations noted*: This is D. M. Ross and L. Sutton, "The Association Between the Hermit Crab *Dardanus arrosor* (Herbst) and the Sea Anemone *Calliactis parasitica* (Couch)," *Proceedings of the Royal Society B* 155, no. 959 (1961): 282–91. See also Graeme D. Ruxton and Martin Stevens, "The Evolutionary Ecology of Decorating Behaviour," *Biology Letters* 11, no. 6 (2015): 20150325, for a good review of body decoration in animals.

  Some anemones even produce a "pseudo-shell" for the crab that grows as the crab does, removing the need for the hermit to change shells. See Hiroki Kise et al., "A Molecular Phylogeny of Carcinoecium-Forming *Epizoanthus* (Hexacorallia: Zoantharia) from the Western Pacific Ocean with Descriptions of Three New Species," *Systematics and Biodiversity* 17, no. 8 (2019): 773–86.

94 *Alongside the stages in animal evolution we have been looking at in the last two chapters*: See Jeremy B. C. Jackson, Leo W. Buss, and Robert E. Cook, eds., *Population Biology and Evolution of Clonal Organisms* (New Haven, CT: Yale University Press, 1985). I had a close look at some cases in an earlier book, *Darwinian Populations and Natural Selection* (Oxford, UK: Oxford University Press, 2009).

95 *Modular organisms often produce branched, tree-like forms*: There are a few branching annelids, which are unitary organisms. See Christopher J. Glasby, Paul C. Schroeder, and María Teresa Aguado, "Branching Out: A Remarkable New Branching Syllid (Annelida) Living in a *Petrosia* Sponge (Porifera: Demospongiae)," *Zoological Journal of the Linnean Society* 164, no. 3 (2012): 481–97: "We describe the morphology and biology of a previously unknown form of branching annelid, *Ramisyllis multicaudata* gen. et sp. nov., an endosymbiont of shallow-water marine sponges (*Petrosia* sp., Demospongiae) in northern Australia. It belongs to the polychaete family Syllidae, as does *Syllis ramosa* McIntosh, 1879, the only other named branching annelid, which was collected from deep-water hexactinellid sponges during the 1875 *Challenger* expedition."

95 *Bryozoans, the shrub-like creatures we saw hosting nudibranchs*: See Matthew H. Dick et al., "The Origin of Ascophoran Bryozoans Was Historically Contingent but Likely," *Proceedings of the Royal Society B* 276 (2009): 3141–48.

100 *During this period I read more about these shrimp*: Victor R. Johnson Jr., "Behavior Associated with Pair Formation in the Banded Shrimp *Stenopus hispidus* (Olivier)," *Pacific Science* 23, no. 1 (1969): 40–50, and Johnson, "Individual Recognition in the Banded Shrimp *Stenopus hispidus* (Olivier)," *Animal Behaviour* 25, pt. 2 (1977): 418–28. See also theaquariumwiki.com/wiki/Stenopus_hispidus.

  Some other crustaceans can recognize individuals. See Joanne Van der Velden et al., "Crayfish Recognize the Faces of Fight Opponents," *PLOS ONE* 3, no. 2

(2008): e1695, and Roy Caldwell's "A Test of Individual Recognition in the Stomatopod *Gonodactylus festate*," *Animal Behaviour* 33, no. 1 (1985): 101–6.

101 *Ascidians coughed and sneezed*: In the first pages of this book I compared some ascidian behaviors to "a shrug and sigh." What are they actually doing? They seem to be clearing out the water held in their bodies. Sometimes one can see waste expelled as well, but often the motion seems to function more as a laconic sort of sneeze. That term, "sneeze," is used by biologists writing about sponges; some sponges deal with murky water with a slow-motion sneeze (Leys, "Elements of a 'Nervous System' in Sponges"). This is another form of coordinated action that might be very old.

Ascidians are notorious as the subjects of an academic quip. It is said that they are mobile when young, and once settled (tenured), they consume their own brain. The anecdote was made well-known by the neuroscientist Rodolfo Llinás, who figures later in this book. George O. Mackie and Paolo Burighel, in the course of a paper about ascidians, reject this vignette with some force and a hint of indignation: "In fact, adult ascidians have perfectly good brains, an order of magnitude larger than those of their larvae, and their behaviour is as finely adapted to sessility as that of the larvae to motility." Mackie and Burighel, "The Nervous System in Adult Tunicates: Current Research Directions," *Canadian Journal of Zoology* 83, no. 1 (2005): 151–83.

5. THE ORIGIN OF SUBJECTS

104 *The philosopher Susan Hurley introduced a good image*: This is in her *Consciousness in Action* (Cambridge, MA: Harvard University Press, 1998), 249. Here is the passage:

> A certain traditional way of thinking about persons still seems natural to many. The core of a person is seen as a subject and an agent standing, so to speak, back to back. The subject has conscious perceptual experience, and is at times self-conscious, in the sense that conscious experience is itself sometimes the object of consciousness (as opposed to worldly objects). The agent makes efforts, tries to do various things, and is at times self-determined, in that certain actions, efforts, or states of will are themselves sometimes of [*sic*—I think she means "an" or "the"?] object of trying (as opposed to worldly events). The subject is the last stop on the input side of the person; the world impinges on the subject. The agent is the first stop on the output side; the agent impinges on the world.

Hurley distinguishes different kinds of entanglement between the sensory and agential side. In some cases, actions and their sensory consequences are distinct from each other—you turn over a rock to see what is underneath. Other situations see a tighter relationship. Your eyes are continually darting around, whether you realize it or not, so the impressions on your retina are continually changing, but this is not registered as change in the scene before you. Humans quite thoroughly adapt to "inverting goggles" that take things on the left into our right field of vision, and vice versa. People come to see things in their proper places—"proper" as determined by what happens when you try to act.

106 *René Descartes, in the seventeenth century, imagined that he might be a soul*: See his *Meditations on First Philosophy* (1641).

106 *A physical duplicate of you might be a mere "zombie," as David Chalmers has put it*:

See his book *The Conscious Mind: In Search of a Fundamental Theory* (Oxford, UK: Oxford University Press, 1996).

107 *Thomas Nagel, despite being a critic of materialism, diagnosed this quirk*: This is in a footnote to his 1974 "What Is It Like to Be a Bat?" paper. See also my "Evolving Across the Explanatory Gap," *Philosophy, Theory, and Practice in Biology* 11, no. 001 (2019): 1–24.

108 *Some critics, like the philosopher Daniel Dennett*: See especially his paper "Quining Qualia," in *Consciousness in Contemporary Science*, ed. Anthony J. Marcel and E. Bisiach (Oxford, UK: Oxford University Press, 1988), 42–77.

109 *Some critiques of materialism seem to want a third-person description*: In "Evolving Across the Explanatory Gap," I say that the whole idea of a "third-person point of view" is a bit confused—all points of view are first-person. But the familiar language helps a bit here.

110 *Hurley brought to philosophy a distinction, used in the psychology and neurobiology of vision, between "what" systems and "where" systems in our brains*: This is also in *Consciousness in Action*.

110–11 *Hurley thought that the sort of processing that a "where" system does for us*: *Consciousness in Action*, 326.

111 *These were the "simple ideas" and "impressions" of the empiricist philosophy of those days*: A paradigm example, also very readable, is Hume's *An Enquiry Concerning Human Understanding* (1748). For the "sense-data" theories that followed in the early twentieth century, see (for example) Bertrand Russell, *The Problems of Philosophy* (New York: Henry Holt, 1912).

112 *The German "idealist" project in philosophy*: This tradition runs especially from Immanuel Kant's *Critique of Pure Reason* (1781) to the work of G. W. F. Hegel (such as his *Phenomenology of Spirit*, 1807). An English-speaking continuation of that tradition often set the situation up as a showdown between more active and more passive views—William James, at the end of the nineteenth century, was an acute observer of this opposition (see *The Will to Believe, and Other Essays in Popular Philosophy*, 1896).

112 *A view called* enactivism *tries, at least in some versions, to explain perception itself as a form of action*: Enactivism now has a large literature, and perhaps not all versions endorse that view. See J. Kevin O'Regan and Alva Noë, "A sensorimotor account of vision and visual consciousness," *Behavioral and Brain Sciences* 24, no. 5 (2001): 939–1031, and Noë's book *Out of Our Heads: Why You Are Not Your Brain, and Other Lessons from the Biology of Consciousness* (New York: Hill and Wang, 2009). Quotes from the former work: "We propose that seeing is a way of acting. It is a particular way of exploring the environment" (from their abstract); "[E]xperiences, we have argued, are not states. They are ways of acting. They are things we do" (p. 960). The term "enactive" in this context was introduced by Francisco J. Varela, Evan Thompson, and Eleanor Rosch in their book *The Embodied Mind: Cognitive Science and Human Experience* (Cambridge, MA: MIT Press, 1991).

112 *As the American philosopher John Dewey wryly noted*: This is in his *Experience and Nature* (Chicago: Open Court, 1925), 36.

113 *Jesse Prinz, who I worked with in New York*: See his book *The Conscious Brain: How Attention Engenders Experience* (Oxford, UK: Oxford University Press, 2012), 341–42. For Dretske, see "Conscious Experience," *Mind* 102, no. 406 (1993): 263–83. Here is Dretske on some problem cases for him:

NOTES

> Why can't we, following Damasio . . . , conceive of emotions, feelings,
> and moods as perception of chemical, hormonal, visceral, and muscuoskel-
> etal states of the body? . . . This way of thinking about pains, itches, tick-
> les, and other bodily sensations puts them in exactly the same category as
> the experiences we have when we are made perceptually aware of our
> environment.

I am indebted to Leonard Katz for discussion of the case of energy level.

115 *Here is a quote from the philosopher John Searle*: This appears in his "Consciousness," *Annual Review of Neuroscience* 23 (2000): 557–78, p. 573.

115 *within neuroscience, in contrast to philosophy and psychology, some prominent figures, including Rodolfo Llinás*: See Llinás and D. Paré, "Of dreaming and wakefulness," *Neuroscience* 44, no. 3 (1991): 521–35. Searle also cites Giulio Tononi and Gerald M. Edelman; see their "Consciousness and Complexity," *Science* 282 (1998): 1846–51.

116 *The idea of a sense of presence plays an uncertain role*: A more careful statement: "[T]he concept of *presence* is used to refer to the subjective sense of reality of the world and of the self within the world." Anil K. Seth, Keisuke Suzuki, and Hugo D. Critchley, "An Interoceptive Predictive Coding Model of Conscious Presence," *Frontiers in Psychology* 2 (2012): 395. See also the quote from Hurley on page 111, where she talks of "a sense of being a self present in the world." And see the quotes from Evan Thompson below.

116 *Some people who accept this also believe a view that is known as* transparency: I have always found transparency to be one of the most peculiar and unlikely ideas in a debate full of peculiar and unlikely ideas. Suppose you deliberately blur your vision a little while looking at something. You will experience blurriness in your visual field. This, surely, is not a sudden attribution of blurriness to the objects you are looking at; it's just a feature of experience itself. That is a good (and well-known) argument against transparency, though replies have been offered and there is some back-and-forth. (The example, now common, might have first been used in Paul A. Boghossian and J. David Velleman, "Colour as a Secondary Quality," *Mind* [n.s.] 98, no. 389 [1989]: 81–103.) An influential defense of transparency is Gilbert Harman, "The Intrinsic Quality of Experience," in *Philosophical Perspectives* 4: *Action Theory and Philosophy of Mind*, ed. James E. Tomberlin (Atascadero, CA: Ridgeview, 1990), 31–52.

116 *Related ideas are sometimes presented by people writing about meditation*: An example is Sam Harris, *Waking Up: A Guide to Spirituality Without Religion* (New York: Simon and Schuster, 2014). The "no-self" doctrine within traditional Buddhism is related, but I don't know enough about Buddhism to make a detailed comparison.

116–17 *Once presence is recognized, it is tempting to see it as an automatic feature*: Something like this view might be expressed in the philosopher Evan Thompson's book *Mind in Life: Biology, Phenomenology, and the Sciences of the Mind* (Cambridge, MA: Harvard University Press, 2010). Thompson suggests that part of the problem of subjective experience is solved by recognizing a basic feeling of animation and presence that comes just from being a living, self-defining system:

> [O]ne might describe consciousness in the sense of sentience as a kind of primitively self-aware liveliness or animation of the body. (p. 161) . . . Earlier in this book I described sentience as the feeling of being alive. Being

sentient means being able to feel the presence of one's body and the world. Sentience is grounded on the autopoietic identity and sense-making of living beings, but in addition it implies a feeling of self and world. (p. 221)

117 *The rubber hand illusion is also the tip of an iceberg*: Here I use Olaf Blanke and Thomas Metzinger, "Full-Body Illusions and Minimal Phenomenal Selfhood," *Trends in Cognitive Sciences* 13, no. 1 (2009): 7–13, and Frédérique de Vignemont's book *Mind the Body: An Exploration of Bodily Self-Awareness* (Oxford, UK: Oxford University Press, 2018).

119–20 *Given all this, it is not mysterious that seeing feels like something*: When I first discussed the interaction between sensing and acting in Chapter 4, I noted that hearing differs markedly from vision and touch in these ways; the effects of your actions on what you hear are real, but often slighter. The roles of "where" systems and cross-checking across the senses are also different. Hearing is as clear a case of sensing as vision is. This is something to keep an eye on.

6. THE OCTOPUS

124 *The decorator crabs, in their protective and deceptive clothing*: I just once came across an octopus eating a decorator crab; I don't know the details of the interaction, as I arrived late in the process.

125 *But soon after the Cambrian, some early cephalopods lifted off*: As in *Other Minds*, for the history I draw on Björn Kröger, Jakob Vinther, and Dirk Fuchs, "Cephalopod Origin and Evolution: A Congruent Picture Emerging from Fossils, Development and Molecules," *BioEssays* 33, no. 8 (2011): 602–13, and a newer paper: Alastair R. Tanner et al., "Molecular Clocks Indicate Turnover and Diversification of Modern Coleoid Cephalopods During the Mesozoic Marine Revolution," *Proceedings of the Royal Society B* 284 (2017): 20162818.

127 *The result, something like 100 million years ago*: A single 165-million-year-old French fossil has been interpreted as an octopus (and was presented this way in *Other Minds*), but further work has suggested that this animal had a hard internal structure, a "gladius" like the sort seen in squid, and hence was not all the way to an octopus design but may have been more similar to a vampire squid; see Isabelle Kruta et al., "*Proteroctopus ribeti* in Coleoid Evolution," *Palaeontology* 59, no. 6 (2016): 767–73.

127 *Some biologists looking closely at insects*: See Gabriella H. Wolf and Nicholas J. Strausfeld, "Genealogical Correspondence of a Forebrain Centre Implies an Executive Brain in the Protostome–Deuterostome Bilaterian Ancestor," *Philosophical Transactions of the Royal Society B* 371 (2016): 20150055.

128 *In 2018, some octopuses were given the drug MDMA*: This paper is Eric Edsinger and Gül Dölen, "A Conserved Role for Serotonergic Neurotransmission in Mediating Social Behavior in Octopus," *Current Biology* 28, no. 19 (2018): 3136–42.e4.

129 *Decades ago, Roger Hanlon and John Messenger suggested*: This is in their *Cephalopod Behaviour*, 1st ed. (Cambridge, UK: Cambridge University Press, 1996).

130 *The lab that has looked most closely at these questions since then is Benny Hochner's*: The papers resulting include several used here: Tamar Gutnick et al., "*Octopus vulgaris* Uses Visual Information to Determine the Location of Its Arm," *Current Biology* 21, no. 6 (2011): 460–62; Letizia Zullo et al., "Nonsomatotopic Organization of the Higher Motor Centers in Octopus," *Current Biology* 19, no. 19 (2009): 1632–36;

and Hochner's "How Nervous Systems Evolve in Relation to Their Embodiment: What We Can Learn from Octopuses and Other Molluscs," *Brain, Behavior, and Evolution* 82 (2013): 19–30.

131 *An important case that is close to home is* lateralization: See Lesley J. Rogers, Giorgio Vallortigara, and Richard Andrew, *Divided Brains: The Biology and Behaviour of Brain Asymmetries* (Cambridge UK: Cambridge University Press, 2013), and Giorgio Vallortigara, Lesley J. Rogers, and Angelo Bisazza, "Possible Evolutionary Origins of Cognitive Brain Lateralization," *Brain Research Reviews* 30 (1999): 164–75.

132 *In cuttlefish, which are cephalopods like octopuses, the right eye is preferred for feeding*: See Alexandra K. Schnell et al., "Lateralization of Eye Use in Cuttlefish: Opposite Direction for Anti-Predatory and Predatory Behaviors," *Frontiers in Physiology* 7 (2016): 620.

133 *A setting where the perplexities and charms of octopus behavior*: This site was described in *Other Minds*. See also the newer paper by David Scheel et al., "Octopus Engineering, Intentional and Inadvertent," *Communicative & Integrative Biology* 11, no. 1 (2018): e1395994. A behavioral paper is Scheel, Godfrey-Smith, and Matthew Lawrence, "Signal Use by Octopuses in Agonistic Interactions," *Current Biology* 26, no. 3 (2016): 377–82.

133 *octopuses have surprisingly short life spans*: There are exceptions, all living in the deep sea. See Chapter 7 of *Other Minds* for these cases.

136 *Alternatively, more sociality might exist in octopus lives*: In "Signal Use by Octopuses in Agonistic Interactions" we listed known exceptions, involving twelve species.

136 *Two other divers, Marty Hing and Kylie Brown, exploring the same general area*: See David Scheel et al., "A Second Site Occupied by *Octopus tetricus* at High Densities, with Notes on Their Ecology and Behavior," *Marine and Freshwater Behaviour and Physiology* 50, no. 4 (2017): 285–91.

137 *A particularly intriguing behavior is throwing*: We are preparing a paper about this at the moment, with the provisional title "Octopuses Throw Debris, Often Hitting Others, with Behavioral Consequences."

140 *This, as Jennifer Mather has noted*: See her "What Is in an Octopus's Mind?," *Animal Sentience* 2019.209.

141 *A famous neurological patient, with the apt name of Ian Waterman*: See Shaun Gallagher's *How the Body Shapes the Mind* (Oxford, UK: Clarendon Press / Oxford University Press, 2005) for discussion.

143 *Bret Grasse, who manages octopuses and other cephalopods*: This was described by Ben Guarino in "Inside the Grand and Sometimes Slimy Plan to Turn Octopuses into Lab Animals," *The Washington Post*, March 3, 2019.

149 *The biologist and roboticist Frank Grasso wrote a paper*: The reference is "The Octopus with Two Brains: How Are Distributed and Central Representations Integrated in the Octopus Central Nervous System?," in *Cephalopod Cognition*, ed. Anne-Sophie Darmaillacq, Ludovic Dickel, and Jennifer Mather (Cambridge, UK: Cambridge University Press, 2014), 94–122. For Sidney Carls-Diamante's discussion, see her "The Octopus and the Unity of Consciousness," *Biology and Philosophy* 32, no. 6 (2017): 1269–87.

149 *The most thorough exploration of the 1+1 possibility I know*: Adrian Tchaikovsky, *Children of Ruin* (New York: Orbit / Hachette, 2019).

150 *We have patients whose bridge between the two brain hemispheres has been cut*: Thomas Nagel's 1971 paper is still a good way into these issues: "Brain Bisection and the Unity

NOTES

of Consciousness," *Synthese* 22, no. 3/4 (1971): 396–413. Tim Bayne's book *The Unity of Consciousness* (Oxford, UK: Oxford University Press, 2010) is a thorough treatment of unity issues. I make extensive use here of Elizabeth Schechter's book, *Self-Consciousness and "Split" Brains: The Minds' I* (Oxford, UK: Oxford University Press, 2018).

151 *I think a version of the fast switching view might be right*: Versions of this view have been defended by several people over the years, sometimes without much detail. Michael Tye's *Consciousness and Persons: Unity and Identity* (Cambridge, MA: MIT Press, 2003) and Adrian Downey's paper discussed below are more detailed. Elizabeth Schechter cites Jerome A. Shaffer's paper "Personal Identity: The Implications of Brain Bisection and Brain Transplants," *The Journal of Medicine and Philosophy* 2, no. 2 (1977): 147–61, as an early statement.

151 *Here I draw on the philosopher Elizabeth Schechter's work*: This is the book cited above, *Self-Consciousness and "Split" Brains.*

152 *Another possibility, again, is that much of the time, both halves of the brain work together*: In some cases, the everyday life of a split-brain patient shows signs of ongoing disunity, as evidenced by disagreement; there are cases where one hand will put on a shirt or get out a cigarette, for example, and the other hand will oppose this action. If that sort of behavior was common, it would be tempting to say that the two-minds condition was permanent (as Schechter thinks it is). Maybe some cases are like this and others have more unity? I take a closer look at these phenomena in "Integration, Lateralization, and Animal Experience" in *Mind and Language.*

153 *Susan Hurley, whose ideas played a role in Chapter 5 of this book*: See her "Action, the Unity of Consciousness, and Vehicle Externalism," in *The Unity of Consciousness: Binding, Integration, and Dissociation*, ed. Axel Cleeremans (Oxford, UK: Oxford University Press, 2003).

153 *Another philosopher, Adrian Downey, recently showed*: See his paper "Split-Brain Syndrome and Extended Perceptual Consciousness," *Phenomenology and the Cognitive Sciences* 17 (2018): 787–811.

153 *The Wada test, named after its inventor, the Japanese-Canadian doctor Juhn Atsushi Wada*: My discussion here draws mostly on James Blackmon's paper, "Hemispherectomies and Independently Conscious Brain Regions," *Journal of Cognition and Neuroethics* 3, no. 4 (2016): 1–26. The quote in the text is from a blog post: jcblackmon.com/general/the-wada-test-for-philosophers-what-is-it-like-to-be-a-proper-part-of-your-own-brain-losing-and-regaining-other-proper-parts-of-your-brain, which in turn uses a public website about epilepsy: epilepsy.com/connect/forums/surgery-and-devices/wada-test-1.

The most extreme version I've encountered of the view that a brain contains many distinct consciousnesses has been defended, in much detail, by the neuroscientist Semir Zeki. For a summary of his position, see his "The Disunity of Consciousness," *Trends in Cognitive Sciences* 7, no. 5 (2003): 214–18.

156 *"Partial unity," again, is the idea that in split-brain cases there is not a neat count of minds*: Schechter wrote an earlier paper about this option, defending its coherence. This is her "Partial Unity of Consciousness: A Preliminary Defense," in *Sensory Integration and the Unity of Consciousness*, ed. David J. Bennett and Christopher S. Hill (Cambridge, MA: MIT Press, 2014), 347–73.

157 *Split-brain cases are full of complexities*: Roger Sperry, in a passage used by Schechter, says that each hemisphere has its own memories, not accessible to the other. (See Sperry's "Hemisphere Deconnection and Unity in Conscious Awareness,"

American Psychologist 23, no. 10 [1968]: 723–33). That would suggest a permanent two-minds situation. But the Sperry paper and other material I've seen seem to be talking about memories laid down in one stage of an experiment, recalled in another stage of the same experiment. It would be different—and perhaps this is how things actually are—if memories remain associated with one hemisphere or the other over the longer term. This would be another reason to think that a two-minds state is permanent.

161 *When I try to imagine this I find myself in a rather hallucinogenic place*: This passage uses some material from my earlier "Octopus Experience," in *Animal Sentience* 2019.270, a commentary on Jennifer Mather's "What Is in an Octopus's Mind?" Incidentally, some echinoderms apparently can *see*—picking out objects—with their whole bodies: see Divya Yerramilli and Sönke Johnsen, "Spatial Vision in the Purple Sea Urchin *Strongylocentrotus purpuratus* (Echinoidea)," *Journal of Experimental Biology* 213, no. 2 (2010): 249–55.

163 *Echinoderms have been around at least since the Cambrian*: Perhaps earlier; see James G. Gehling, "Earliest Known Echinoderm—A New Ediacaran Fossil from the Pound Subgroup of South Australia," *Alcheringa* 11, no. 4 (1987): 337–45. See Samuel Zamora, Imran A. Rahman, and Andrew B. Smith, "Plated Cambrian Bilaterians Reveal the Earliest Stages of Echinoderm Evolution," *PLOS One* 7, no. 6 (2012): e38296, for a discussion of their path away from the usual bilaterian plan, with some very good images. I think the species I was seeing near Octopolis was *Antedon loveni*.

7. KINGFISH

166 *In a book by the biologist Neil Shubin*: This is *Your Inner Fish: A Journey into the 3.5-Billion-Year History of the Human Body* (New York: Pantheon, 2008).

166 *Fish began very much as minor players*: My main source through here is John A. Long's *The Rise of Fishes: 500 Million Years of Evolution* (Baltimore: Johns Hopkins University Press, 1995). Long's book is rather full of talk of evolutionary ladders and scales, something I criticize from time to time here. A good book that covers early vertebrate evolution with an eye specifically on consciousness is Todd E. Feinberg and Jon M. Mallatt, *The Ancient Origins of Consciousness: How the Brain Created Experience* (Cambridge, MA: MIT Press, 2016). A recent paper argues for the role of nearshore environments in early fish evolution: Lauren Sallan et al., "The Nearshore Cradle of Early Vertebrate Diversification," *Science* 362 (2018): 460–64 (another paper with good images).

167 *In a classic example of what the French biologist François Jacob has described*: See his "Evolution and Tinkering," *Science* 196 (1977): 1161–66.

168 *In 2018, an amateur fossil hunter in Australia*: This was reported by SBS News (sbs .com.au/news/rare-set-of-mega-shark-teeth-from-prehistoric-species-unearthed): "Think a Steak Knife. They're Sharp." *Carcharocles angustidens* grew to more than nine meters.

169 *feeling magnetically the whole of Earth*: See Sönke Johnsen and Kenneth J. Lohmann, "The Physics and Neurobiology of Magnetoreception," *Nature Reviews Neuroscience* 6 (2005): 703–12.

169 *The whale shark is the biggest fish in the oceans today*: Probably not the biggest ever; that seems to have been the dinosaur-era *Leedsichthys problematicus*.

169 *At the time of writing, it is suspected that they all mate at a particular place*: Here I am drawing on the BBC TV show *Blue Planet II* (2017).

A note on the sex of the one I was following: on Ningaloo Reef, the sex ratio is male-biased by about three to one. The reason for this is not known.

170 *The transition to bonier vertebrates may have come with the duplication*: See Darja Obradovic Wagner and Per Aspenberg, "Where Did Bone Come From?," *Acta Orthopaedica* 82, no. 4 (2011): 393–98.

171 *This is a tactile sense, roughly speaking, a form of touch*: I make much use of Horst Bleckmann and Randy Zelick, "Lateral Line System of Fish," *Integrative Zoology* 4, no. 1 (2009): 13–25. For lateral line obsessives, there is a book: Sheryl Coombs et al., eds., *The Lateral Line System* (New York: Springer, 2014). For the "long-distance touch" description, see John Montgomery, Horst Bleckmann, and Sheryl Coombs, "Sensory Ecology and Neuroethology of the Lateral Line," in *The Lateral Line System*, 121–50.

172 *The evolutionary story goes something like this*: Here I use Bernd Fritzsch and Hans Straka, "Evolution of Vertebrate Mechanosensory Hair Cells and Inner Ears: Toward Identifying Stimuli That Select Mutation Driven Altered Morphologies," *Journal of Comparative Physiology A* 200, no. 1 (2014): 5–18; Bernd U. Budelmann and Horst Bleckmann, "A Lateral Line Analogue in Cephalopods: Water Waves Generate Microphonic Potentials in the Epidermal Head Lines of *Sepia* and *Lolliguncula*," *Journal of Comparative Physiology A* 164, no. 1 (1988): 1–5.

173 *Even a small fish leaves a wake*: Bleckmann and Zelick, "Lateral Line System of Fish." Lateral line sensing has rich interactions between sensing and action: John C. Montgomery and David Bodznick, "An Adaptive Filter That Cancels Self-Induced Noise in the Electrosensory and Lateral Line Mechanosensory Systems of Fish," *Neuroscience Letters* 174, no. 2 (1994): 145–48.

173 *Blind cave fish, as the name suggests*: Bleckmann and Zelick, "Lateral Line System of Fish." These fish use lateral line information to build internal maps of their environments: Theresa Burt de Perera, "Spatial Parameters Encoded in the Spatial Map of the Blind Mexican Cave Fish, *Astyanax fasciatus*," *Animal Behaviour* 68, no. 2 (2004): 291–95.

174 *In some fish, especially sharks, the lateral line system has been modified*: Clare V. H. Baker, Melinda S. Modrell, and J. Andrew Gillis, "The Evolution and Development of Vertebrate Lateral Line Electroreceptors," *The Journal of Experimental Biology* 216, pt. 13 (2013): 2515–22, and Nathaniel B. Sawtell, Alan Williams, and Curtis C. Bell, "From Sparks to Spikes: Information Processing in the Electrosensory Systems of Fish," *Current Opinion in Neurobiology* 15, no. 4 (2005): 437–43. The only terrestrial animals that have electrosensing seem to be monotremes (platypus, echidna): John D. Pettigrew, "Electroreception in Monotremes," *Journal of Experimental Biology* 202 (1999): 1447–54.

174 *The late shark researcher Aidan Martin described seeing hammerheads*: See elasmo-research.org/education/topics/d_functions_of_hammer.htm.

174 *catfish, whose form of electrosensing seems to be so sensitive that it predicts earthquakes*: "Because of its general interest, something must be said about earthquake prediction by catfish. This is so far the best documented case of animal responses prior to the major shock. The series of papers from Japan in the early 1930s that established a statistically significant behavior change some hours before an earthquake are reviewed by Kalmijn (1974). The Japanese authors gave some evidence that the adequate stimulus was electric potentials from Earth. Only later quantitative measurements of catfish sensitivity and of the magnitude of potentials anticipating

the main quake (which are not universal accompaniments of earthquakes and are perhaps peculiar to certain regions) showed the signals to be well above threshold and the original observations therefore quite plausible." Theodore H. Bullock, "Electroreception," *Annual Review of Neuroscience* 5 (1982): 121–70, p. 128.

175 *What surprises me is that research looking directly at the role of the lateral line system in schooling is equivocal*: See, for example, Prasong J. Mekdara et al., "The Effects of Lateral Line Ablation and Regeneration in Schooling Giant Danios," *Journal of Experimental Biology* 221 (2018): jeb175166.

175–76 *I came across unusual numbers of a fish I'd seen there often*: The smooth flutemouth, *Fistularia commersonii*.

177 *Somewhere in this sequence, some fish also became smart*: In this section I draw on Redouan Bshary and Culum Brown, "Fish Cognition," *Current Biology* 24, no. 19 (2014): R947–50, and references that come from that paper. Also Jonathan Balcombe, *What a Fish Knows: The Inner Lives of Our Underwater Cousins* (New York: Scientific American / Farrar, Straus and Giroux, 2016).

For counting, see Christian Agrillo et al., "Use of Number by Fish," *PLOS One* 4, no. 3 (2009): e4786. The music recognition experiments are described in Balcombe's book.

178 *This principle was originally developed for primates*: See Nicholas K. Humphrey, "The Social Function of Intellect," in *Growing Points in Ethology*, ed. P. P. G. Bateson and R. A. Hinde (Cambridge, UK: Cambridge University Press, 1976), 303–17.

179 *Most known fish hang around with other fish*: Matz Larsson, "Why Do Fish School?," *Current Zoology* 58, no. 1 (2012): 116–28. I wonder whether the yellow vertical bands that quite a few silver fish have on their tails, including the kingfish and others, function in schooling—giving a visual cue of location and movement. This would require that the individual fish benefits from being visible and predictable. It is not enough for it to make things easier for watcher fish, or for the school as a whole.

179 *Jean-Paul Sartre (who, incidentally, had an intense and drug-amplified fear*: See, among other works, Thomas Riedlinger, "Sartre's Rite of Passage," *Journal of Transpersonal Psychology* 14, no. 2 (1982): 105–23.

179 *The cleverness of fish shows up especially in their dealing with others*: See Jeremy R. Kendal et al., "Nine-Spined Sticklebacks Deploy a Hill-Climbing Social Learning Strategy," *Behavioral Ecology* 20, no. 2 (2009): 238–44; Stefan Schuster et al., "Animal Cognition: How Archer Fish Learn to Down Rapidly Moving Targets," *Current Biology* 16, no. 4 (2006): 378–83; Logan Grosenick, Trisha S. Clement, and Russell D. Fernald, "Fish Can Infer Social Rank by Observation Alone," *Nature* 445 (2007): 429–32.

180 *One kind of cleaner fish tends to cheat less if there are onlookers*: See Ana Pinto et al., "Cleaner Wrasses *Labroides dimidiatus* Are More Cooperative in the Presence of an Audience," *Current Biology* 21, no. 13 (2011): 1140–44.

180 *On many seafloors you may find shrimp-goby collaborations*: There are many papers about these associations, e.g., Annemarie Kramer, James L. Van Tassell, and Robert A. Patzner, "A Comparative Study of Two Goby Shrimp Associations in the Caribbean Sea," *Symbiosis* 49 (2009): 137–141.

181 *These fish hunt cooperatively with moray eels*: Alexander L. Vail, Andrea Manica,

and Redouan Bshary, "Referential Gestures in Fish Collaborative Hunting," *Nature Communcations* 4 (2013): 1765.

182 *Hans Berger, in his laboratory in Germany in the early 1920s*: I use David Millett's "Hans Berger: From Psychic Energy to the EEG," *Perspectives in Biology and Medicine* 44, no. 4 (2001): 522–42. Some tellings of the Berger story seem to have become a little distorted.

183 *A younger colleague, Raphael Ginzberg, who became a subject of some of Berger's experiments, described him later*: This is described in Millett's "Hans Berger: From Psychic Energy to the EEG."

184 *He was not the first to do so*: Vladimir Vladimirovich Pravdich-Neminsky is apparently the precursor in this case. He was arrested by the Soviet regime and was able to do little work afterward.

After his death, Berger was for some time seen as an opponent of the Nazis, but recent work suggests he was not at all. See Lawrence A. Zeidman, James Stone, and Daniel Kondziella, "New Revelations About Hans Berger, Father of the Electroencephalogram (EEG), and His Ties to the Third Reich," *Journal of Child Neurology* 29, no. 7 (2014): 1002–10.

185 *This is a result of a duality in the role of charge*: Here I am not talking about the duality of electric and magnetic, in electromagnetic phenomena; this is a more informal use of the term. In this chapter I am setting aside the magnetic aspect as much as I can.

186 *The patterns in an EEG are mostly due to those slower changes*: Here, and through much of this initial discussion of rhythms and fields, I use György Buzsáki, *Rhythms of the Brain* (Oxford, UK: Oxford University Press, 2006), and (a much more technical but still helpful presentation) György Buzsáki, Costas A. Anastassiou, and Christof Koch, "The Origin of Extracellular Fields and Currents—EEG, ECoG, LFP and Spikes," *Nature Reviews Neuroscience* 13 (2012): 407–20. The point mentioned in the italicized sentence—the idea that EEG patterns are mostly not due to spikes—is contentious and varies by context and by the EEG pattern in question; see the review above.

186 *When this is done with fruit flies, crayfish, octopuses, and many other animals*: Some examples: Theodore H. Bullock, "Ongoing Compound Field Potentials from Octopus Brain Are Labile and Vertebrate-Like," *Electroencephalography and Clinical Neurophysiology* 57, no. 5 (1984): 473–83; R. Aoki et al., "Recording and Spectrum Analysis of the Planarian Electroencephalogram," *Neuroscience* 159, no. 2 (2009): 908–14; Bruno van Swinderen and Ralph J. Greenspan, "Salience Modulates 20–30 Hz Brain Activity in *Drosophila*," *Nature Neuroscience* 6 (2003): 579–86; Fidel Ramón et al., "Slow Wave Sleep in Crayfish," *Proceedings of the National Academy of Sciences USA* 101, no. 32 (2004): 11857–61.

187 *The idea that the synchronization of activity is an important part of how brains do things*: See György Buzsáki, *Rhythms of the Brain*; Rodolfo R. Llinás, *I of the Vortex: From Neurons to Self* (Cambridge, MA: MIT Press, 2001); Wolf Singer, "Neuronal Oscillations: Unavoidable and Useful?," *European Journal of Neuroscience* 48, no. 7 (2018): 2389–98; Conrado A. Bosman, Carien S. Lansink, and Cyriel M. A. Pennartz, "Functions of Gamma-Band Synchronization in Cognition: From Single Circuits to Functional Diversity Across Cortical and Subcortical Systems," *European Journal of Neuroscience* 39, no. 11 (2014): 1982–99.

188 *Angélique Arvanitaki was a French neurophysiologist*: She does not seem well-known

at all. I used François Clarac and Edouard Pearlstein, "Invertebrate Preparations and Their Contribution to Neurobiology in the Second Half of the 20th Century," *Brain Research Reviews* 54, no. 1 (2007): 113–61.

188 *one is reminded of Barbara McClintock of "jumping genes" and Lynn Margulis of the symbiotic origin of mitochondria*: McClintock eventually won a Nobel Prize. Margulis's role is discussed in a book cited in Chapter 2, John Archibald's *One Plus One Equals One*. For McClintock, see Evelyn Fox Keller, *A Feeling for the Organism: The Life and Work of Barbara McClintock* (New York: Henry Holt, 1983).

188 *Arvanitaki began her most important paper*: This is "Effects Evoked in an Axon by the Activity of a Contiguous One," *Journal of Neurophysiology* 5, no. 2 (1942): 89–108. The work was done in cuttlefish.

189 *In some cases, the rhythmic patterns in fields generated by a whole region of the brain*: The main papers I use through here are Costas A. Anastassiou et al., "Ephaptic Coupling of Cortical Neurons," *Nature Neuroscience* 14, no. 2 (2011): 217–23; Chia-Chu Chiang et al., "Slow Periodic Activity in the Longitudinal Hippocampal Slice Can Self-Propagate Non-Synaptically by a Mechanism Consistent with Ephaptic Coupling," *Journal of Physiology* 597, no. 1 (2019): 249–69; Costas A. Anastassiou and Christof Koch, "Ephaptic Coupling to Endogenous Electric Field Activity: Why Bother?," *Current Opinion in Neurobiology* 31 (2015): 95–103.

189 *As Christof Koch and Costas Anastassiou say, this is a novel feedback mechanism*: In "Ephaptic Coupling to Endogenous Electric Field Activity: Why Bother?"

189 *In a book from a couple of decades ago, Llinás introduces*: This is his *I of the Vortex: From Neurons to Self*.

In his "Review of György Buzsáki's book *Rhythms of the Brain*," *Neuroscience* 149 (2007): 726–27, Llinás opts for the second option of the list I give in the text— the idea that synchronized activity is important but fields and their oscillations are not—but he also adds a qualifier:

> Amazingly, some authors view such rhythms as "emanating" from the brain and as being the ultimate expression of brain function. This view is as absurd as to regard the extracellularly recorded electric fields [as] the biologically significant aspect of nerve conduction. Indeed, except in cases such as the inhibitory effect of VIIIth nerve activity on Mauthner cell's axon in teleosts, or perhaps in some cases of "ephaptic modulation of nerve excitability," *the extracellular field potentials are epiphenomena.* They may report to the external observer the presence of electrical coherence of neuronal groups, but themselves are no more than shadows in a platonic cave.

The message of the more recent work is that that might be a big "except."

190 *I did some reading about why cicadas and their relatives sing*: For example, see M. Hartbauer et al., "Competition and Cooperation in a Synchronous Bushcricket Chorus," *Royal Society Open Science* 1, no. 2 (2014): 140167.

191 *Christiaan Huygens was a seventeenth-century scientific polymath*: Wolf Singer, in his "Neuronal Oscillations: Unavoidable and Useful?," refers to Huygens as "a Dutch watch maker"; this seems a bit unfair to the discoverer of the rings of Saturn.

194 *That analogy is nearly as old as telephone exchanges themselves*: This is in the second edition of Karl Pearson's influential book *The Grammar of Science* (London: Adam and Charles Black, 1900).

195 *The work I have been describing here in this chapter*: All through here I have been

influenced by the work of Rosa Cao, and many years of discussion with her; see her "Why Computation Isn't Enough: Essays in Neuroscience and the Philosophy of Mind" (PhD dissertation, New York University, 2018).

196 *The idea is not that experience or consciousness* is *a field, in the physical sense*: Views of this kind have been proposed by several people—see Susan Pockett, *The Nature of Consciousness: A Hypothesis* (New York: iUniverse, 2000); E. R. John, "A Field Theory of Consciousness," *Consciousness and Cognition* 10 (2001): 184–213; and Johnjoe McFadden, "Synchronous Firing and Its Influence on the Brain's Electromagnetic Field: Evidence for an Electromagnetic Field Theory of Consciousness," *Journal of Consciousness Studies* 9, no. 4 (2002): 23–50. Back in Chapter 5, I used a quote from John Searle to introduce some ideas about what experience tends to include, ideas that get us away from an exclusive focus on sensing. Searle himself calls his approach to consciousness a "unified field" view. I think there is no appeal here to the physical sense of "field." The main role of the field idea in Searle's paper is to emphasize the way that many aspects of experience are integrated into a whole. In the last chapter I use the phrase "experiential profile" to capture ideas like this without bringing in fields.

196 *Some years ago, a few scientists and philosophers raised the idea that a particular kind of high-frequency wave pattern*: See Francis Crick and Christof Koch, "Towards a Neurobiological Theory of Consciousness," *Seminars in the Neurosciences* 2 (1990): 263–75. Of the others making use of this idea, I found Jesse Prinz's "Attention, Working Memory, and Animal Consciousness," in *The Routledge Handbook of Philosophy of Animal Minds*, ed. Kristin Andrews and Jacob Beck (New York: Routledge, 2018), particularly thought-provoking here. The work on gamma was an important innovation, but my view is broader and does not insist on the importance of 40-hertz rhythms.

196 *For example, he and Ralph Greenspan found that a particular wave pattern*: See van Swinderen and Greenspan, "Salience Modulates 20–30 Hz Brain Activity in *Drosophila*."

197 *Suppose we switch to imagining this process in the way that various neurobiologists discussed in this chapter would like us to imagine it*: They probably disagree on many things (Llinás, Buzsáki, Singer, Koch . . .) but probably agree on this point.

199–200 *If you imagine an organism doing none of those things, it won't have the large-scale dynamic properties either*: As Rosa Cao pointed out, an interesting case to think about here is artifically grown tiny collections of neurons—"brainoids." They can exhibit some of the patterns seen in nervous systems.

200 *Similarly, a conscious mind is not just a system whose activities are unusually closely tied together*: The "Integrated Information Theory" (IIT) holds that a high degree of integration of activity in a system of any kind makes that system conscious. (See Giulio Tononi and Christof Koch, "Consciousness: Here, There and Everywhere?," *Philosophical Transactions of the Royal Society B* 370: 20140167.) Integration is important because of its links to *other* properties, properties that have a bridging role in relation to mind and matter: being a subject and occupying a perspective.

200 *Rodolfo Llinás, the neuroscientist who has come up several times in this chapter*: Llinás and Paré, "Of Dreaming and Wakefulness," *Neuroscience* 44, no. 3 (1991): 521–35. "We propose here, as we have done on past occasions . . . that consciousness, like locomotion, might be more a case of intrinsic activity than of sensory drive. Thus, it has been proposed that consciousness is an oneiric-like internal functional state modulated, rather than generated, by the senses."

200 *Merker thinks, instead, that these rhythms might be important in the background upkeep*

NOTES

of brain activity: Björn Merker, "Cortical Gamma Oscillations: The Functional Key Is Activation, Not Cognition," *Neuroscience and Biobehavioral Reviews* 37, no. 3 (2013): 401–17.

8. ON LAND

204   *The first animals to creep up into the sun were arthropods*: See Jason A. Dunlop, Gerhard Scholtz, and Paul A. Selden, "Water-to-Land Transitions," in *Arthropod Biology and Evolution: Molecules, Development, Morphology*, ed. Alessandro Minelli, Geoffrey Boxshall, and Giuseppe Fusco (Berlin: Springer-Verlag, 2013), 417–40. Also Casey W. Dunn, "Evolution: Out of the Ocean," *Current Biology* 23, no. 6 (2013): R241–43.

204   *The land is difficult, and arthropods had features that helped*: Another problem I'll just mention in passing here is UV radiation. This is discussed in George McGhee Jr., *When the Invasion of Land Failed: The Legacy of the Devonian Extinctions* (New York: Columbia University Press, 2013).

204   *Arthropods have made something like seven, perhaps more, separate moves*: Dunlop, Scholtz, and Selden, "Water-to-Land Transitions."

205   *Land plants, especially flowering plants, consume the sun's energy*: See Richard K. Grosberg, Geerat J. Vermeij, and Peter C. Wainwright, "Biodiversity in Water and on Land," *Current Biology* 22, no. 21 (2012): R900–903.

205   *Some of these achievements are so intricate*: See Scarlett R. Howard et al., "Numerical Cognition in Honeybees Enables Addition and Subtraction," *Science Advances* 5, no. 2 (2019): eaav0961; Aurore Avarguès-Weber et al., "Simultaneous Mastering of Two Abstract Concepts by the Miniature Brain of Bees," *Proceedings of the National Academy of Sciences USA* 109, no. 19 (2012): 7481–86; and Olli Loukola et al., "Bumblebees Show Cognitive Flexibility by Improving on an Observed Complex Behavior," *Science* 355 (2017): 833–36.

206   *Bees reserve perhaps their most impressive and also beautiful behaviors*: See Vincent Gallo and Lars Chittka, "Cognitive Aspects of Comb-Building in the Honeybee?," *Frontiers in Psychology* 9 (2018): 900.

207   *They do scout new options among all this*: Much was learned about this mix of efficient use of resources and ongoing exploration of options in Joseph L. Woodgate et al., "Life-Long Radar Tracking of Bumblebees," *PLOS ONE* 11, no. 8 (2016): e0160333.

207   *A few years ago, Jonas Richter and collaborators presented octopuses with a puzzle box*: See Jonas N. Richter, Binyamin Hochner, and Michael J. Kuba, "Pull or Push? Octopuses Solve a Puzzle Problem," *PLOS ONE* 11, no. 3 (2016): e0152048.

208   *The question of a conscious insect is difficult*: For a particularly good treatment of the idea of insect consciousness, see Andrew B. Barron and Colin Klein, "What Insects Can Tell Us About the Origins of Consciousness," *Proceedings of the National Academy of Sciences USA* 113, no. 18 (2016): 4900–908.

208   *Many insects are impressive on the sensory side of things*: I don't discuss spiders much in this book—simply because the cast of characters is so large already—but some spiders are very impressive indeed on the sensory side. These tend to be spiders that don't build webs, but wander and hunt. Jumping spiders (Salticidae) are especially remarkable. See Robert R. Jackson and Fiona R. Cross, "Spider Cognition," *Advances in Insect Physiology* 41 (2011): 115–74.

209 *Some decades ago, Craig Eisemann and a group of colleagues*: Craig H. Eisemann et al., "Do Insects Feel Pain?—A Biological View," *Experientia* 40 (1984): 164–67.

210 *The idea of distinct sensory and evaluative sides to experience*: A very interesting paper by Justin Sytsma and Edouard Machery, "Two Conceptions of Subjective Experience," *Philosophical Studies* 151, no. 2 (2010): 299–327, looks at whether ordinary people recognize a single concept of "felt" or subjective experience in the way philosophers do. Sytsma and Machery find that they don't, and tie experience closely to valuation, treating purely sensory events (seeing red, etc.) in different and thinner ways. Sytsma and Machery may be right about the everyday conception of experience, but everyday thinking may also be mistaken.

Feinberg and Mallatt, in *Ancient Origins of Consciousness*, distinguish three kinds of consciousness: sensory (exteroceptive), affective, and interoceptive.

211 *As a result, people look for markers of something more*: See Lynne U. Snedden et al., "Defining and Assessing Animal Pain," *Animal Behaviour* 97 (2014): 201–12.

212 *Julia Groening and her colleagues*: Julia Groening, Dustin Venini, and Mandyam V. Srinivasan, "In Search of Evidence for the Experience of Pain in Honeybees: A Self-Administration Study," *Scientific Reports* 7 (2017): 45825.

212 *Terry Walters of the University of Texas*: See Edgar T. Walters, "Nociceptive Biology of Molluscs and Arthropods: Evolutionary Clues About Functions and Mechanisms Potentially Related to Pain," *Frontiers in Physiology* 9 (2018): 1049, and Robyn J. Crook and E. T. Walters, "Nociceptive Behavior and Physiology of Molluscs: Animal Welfare Implications," *ILAR Journal* 52, no. 2 (2011): 185–95.

212–13 *A similar emotion-like state was seen by Melissa Bateson and her colleagues*: See Melissa Bateson et al., "Agitated Honeybees Exhibit Pessimistic Cognitive Biases," *Current Biology* 21, no. 12 (2011): 1070–73. The paper about optimism is Clint Perry, Luigi Baciadonna, and Lars Chittka, "Unexpected Rewards Induce Dopamine-Dependent Positive Emotion–Like State Changes in Bumblebees," *Science* 353 (2016): 1529–31, with Solvi as first author, publishing under the name Clint Perry.

213 *Some other discussions of evaluative experience are organized around a distinction*: In Simona Ginsburg and Eva Jablonka's book, *The Evolution of the Sensitive Soul: Learning and the Origins of Consciousness* (Cambridge, MA: MIT Press, 2019), the evolution of "unlimited associative learning" is seen as a transition that marks that appearence of consciousness. I agree that this kind of learning was a very important invention, but do not yet see the intrinsic connection to consciousness, in part because of the work on emotion-like states in simpler animals.

215 *A scene that illustrates this possibility in gastropods*: Winkworth has a good YouTube channel: youtube.com/user/swinkworth.

216 *I was reminded of a comment by Daniel Dennett*: This is in his "Review of *Other Minds*," *Biology & Philosophy* 34, no. 1 (2019): 2.

217 *Sharks and rays, unlike other fish, seem to lack nociceptors*: See Michael Tye's book *Tense Bees and Shell-Shocked Crabs: Are Animals Conscious?* (Oxford, UK: Oxford University Press, 2016). See also Lynne U. Sneddon, "Nociception," *Fish Physiology* 25 (2006): 153–78.

218 *A nice (in both senses) experiment by Marta Soares*: See Marta Soares et al., "Tactile Stimulation Lowers Stress in Fish," *Nature Communications* 2 (2011): 534.

218 *As Robyn Crook and Terry Walters note in a review article*: "Nociceptive Behavior and Physiology of Molluscs: Animal Welfare Implications."

219 *After seeing marks of sentience in insects, snails, and the like*: The insect question has

yet another complication. I have been writing so far about "the insect," the life of the whole animal. But part of the insect adventurousness in lifestyle is *metamorphosis*, the transition from larva to adult, as in caterpillar and butterfly. Many insects lead two lives, in effect, one on each side of a metamorphic divide. Extensive breakdown and reconstruction of the body takes place along the way.

In the kinds of insects considered here, it is the adult who has acute sensing and complex motion; the larva does not. (Larvae in many cases do have eyes, but much simpler ones.) On the other side, more sensitivity to damage is seen in the larvae. Adults have the capacity to tend and protect wounds, but don't. A larva might be more sensitive, but probably *can't* tend its injuries, even if it wanted to. On the other hand, the emotion-like states uncovered in the Bateson and Perry studies are found in adults, and I don't know whether the behaviors of larvae are complicated enough to support optimism and pessimism.

220 *Eventually, some green algae made their way*: Karl J. Niklas's book *The Evolutionary Biology of Plants* (Chicago: University of Chicago Press, 1997) is a few years old but very interesting. For updates on early stages, see Charles H. Wellman, "The Invasion of the Land by Plants: When and Where?," *New Phytologist* 188, no. 2 (2010): 306–309, and Jennifer L. Morris et al., "The Timescale of Early Land Plant Evolution," *Proceedings of the National Academy of Sciences USA* 115, no. 10 (2018): E2274–83.

For a review of animals who photosynthesize in various ways, see Mary E. Rumpho et al., "The Making of a Photosynthetic Animal," *Journal of Experimental Biology* 214 (2011): 303–311. Remnants of mobility do remain in many plants, in their sex cells. Ferns and cycads, among others, have mobile sperm cells that swim through water to achieve fertilization. One tiny part of the life cycle remains mobile. In achieving a more mobile life than this, plants would also be constrained by their thick cell walls; this is something they would have to overcome.

221 *If you also ask plant people which plants are the smartest*: One plant person I ask is Monica Gagliano, author of *Thus Spoke the Plant* (Berkeley, CA: North Atlantic Books, 2018). She has argued that plants can exhibit simple forms of learning; see especially Monica Gagliano et al., "Learning by Association in Plants," *Scientific Reports* 6 (2016): 38427. The plant in question is a climber.

Climbing vines are mostly angiosperms, flowering plants, with a few exceptions (*Gnetum*, a gymnosperm, and *Lygodium*, a fern).

221 *Darwin realized this; he said that down here*: See *The Power of Movement of Plants* (London: John Murray, 1880), which Charles wrote with his son Francis.

222 *I will describe an example that surprised me*: This is in Masatsugu Toyota et al., "Glutamate Triggers Long-Distance, Calcium-Based Plant Defense Signaling," *Science* 361 (2018): 1112–15.

223 *In the case of plants, this was recognized*: See Johann Wolfgang von Goethe's *Metamorphosis of Plants* (1790) and Erasmus Darwin's *Phytologia* (1800).

223 *In special cases, connections might get so tight*: It is interesting to think about the nervous systems of modular animals like corals and bryozoans, in this connection—each "zoid" does have its own nervous system.

224 *I don't agree with the view that plants are just "very slow animals"*: The quote is from a presentation and a BBC story about plant sensing, "Plants Can See, Hear and Smell—and Respond," *BBC Earth*, January 10, 2017.

NOTES

224 *Further electro-botanical surprises may be waiting*: This sort of thing is intriguing: Gabriel R. A. de Toledo et al., "Plant Electrome: The Electrical Dimension of Plant Life," *Theoretical and Experimental Plant Physiology* 31 (2019): 21–46.

225 *As work on sensing and signaling in organisms like plants has progressed, the term "minimal cognition"*: The term is quite controversial, partly for reasons expressed in the title of this article by Pamela Lyon, "Of What Is 'Minimal Cognition' the Half-Baked Version?," *Adaptive Behavior*, September 2019. See also Jules Smith-Ferguson and Madeleine Beekman, "Who Needs a Brain? Slime Moulds, Behavioural Ecology and Minimal Cognition," *Adaptive Behavior*, January 2019.

9. FINS, LEGS, WINGS

229 *The vertebrate story was different*: Here I use Miriam Ashley-Ross et al., "Vertebrate Land Invasions—Past, Present, and Future: An Introduction to the Symposium," *Integrative and Comparative Biology* 53, no. 2 (2013): 192–96, and Jennifer A. Clack, *Gaining Ground: The Origin and Evolution of Tetrapods*, 2nd ed. (Bloomington: University of Indiana Press, 2012).

229 *Dry land for vertebrates is, indeed, a litany of obstructions*: For the catfish swallowing, see Sam Van Wassenbergh, "Kinematics of Terrestrial Capture of Prey by the Eel-Catfish *Channallabes apus*," *Integrative and Comparative Biology* 53, no. 2 (2013): 258–68.

230–31 *Another challenge for vertebrates on land was the unsuitability of their eggs*: This has often been seen as *the* obstacle. On the other hand, recent work has suggested that reproducing on land can't be quite as bad a problem as it seemed, as some present-day fish who don't normally leave the water make these excursions to deposit eggs. Some leave their eggs for a brief land sojourn between tides. In another fish, the splash tetra, mating takes place in the middle of a spectacular tandem leap to an overhanging leaf, where the eggs are left. The male then splashes the eggs every few minutes to keep them moist while they incubate, and the newly hatched fish then drop back into the water.

The land is probably a good place for eggs in these cases because of the abundance of oxygen, and perhaps higher temperatures. Land has a mixture of advantages and disadvantages. See Karen L. M. Martin and A. L. Carter, "Brave New Propagules: Terrestrial Embryos in Anamniotic Eggs," *Integrative and Comparative Biology* 53, no. 2 (2013): 233–47.

231 *Initially, a group called the Synapsids was larger, more numerous, more diverse*: Through here, I make extensive use of Steve Brusatte's *The Rise and Fall of the Dinosaurs: A New History of Their Lost World* (New York: William Morrow, 2018).

235 *Endothermy in its full-fledged form evolved independently in mammals and birds*: Maybe also in proto-mammals or stem-group mammals before the Triassic; this is debated.

There are two distinctions to keep an eye on here. *Endothermy* versus *ectothermy* is producing one's own heat versus gaining it from the environment, respectively; *homeothermy* versus *poikilothermy* is a matter of constant versus variable temperature, respectively. We mammals are both endothermic and homeothermic. A helpful article, discussing a classic earlier paper, is Michael S. Hedrick and Stanley S. Hillman, "What Drove the Evolution of Endothermy?," *Journal of Experimental Biology* 219 (2016): 300–301.

235 *Some meticulous work from Simon Laughlin's laboratory at Cambridge*: See Benjamin

W. Tatler, David O'Carroll, and Simon B. Laughlin, "Temperature and the Temporal Resolving Power of Fly Photoreceptors," *Journal of Comparative Physiology A* 186, no. 4 (2000): 399–407.

235 *In the sea, warm-bloodedness is rare*: See Barbara A. Block et al., "Evolution of Endothermy in Fish: Mapping Physiological Traits on a Molecular Phylogeny," *Science* 260 (1993): 210–14, and Kerstin A. Fritsches, Richard W. Brill, and Eric J. Warrant, "Warm Eyes Provide Superior Vision in Swordfishes," *Current Biology* 15, no. 1 (2005): 55–58. I am grateful to Bill Blessing for prompting me to think about this issue.

236 *Long ago, some predatory marine reptiles—ichthyosaurs and others*: See Jorge Cubo et al., "Bone Histology of *Azendohsaurus laaroussii*: Implications for the Evolution of Thermometabolism in Archosauromorpha," *Paleobiology* 45, no. 2 (2019): 317–30.

236 *Body temperature in dinosaurs is a hotly debated topic*: See Brusatte's *The Rise and Fall of the Dinosaurs*.

238 *The two sides of these brains also seem to use somewhat different "styles" of processing*: Here I make use of a number of papers by Giorgio Vallortigara and Lesley Rogers. (Many thanks to Lesley Rogers for help with this material.) The papers include Giorgio Vallortigara, "Comparative Neuropsychology of the Dual Brain: A Stroll through Animals' Left and Right Perceptual Worlds," *Brain and Language* 73, no. 2 (2000): 189–219, and Lesley J. Rogers, "A Matter of Degree: Strength of Brain Asymmetry and Behaviour," *Symmetry* 9, no. 4 (2017): 57. For a review, see, again, Rogers, Vallortigara, and Andrew, *Divided Brains: The Biology and Behaviour of Brain Asymmetries*.

238 *For example, Giorgio Vallortigara and Luca Tommasi used temporary eye*: See Vallortigara, "Comparative Neuropsychology of the Dual Brain." He also summarizes the toad work.

239 *researchers on lizards and fish, in particular, have found themselves explicitly comparing these animals to "split-brain" cases*: From the Vallortigara paper cited above:

> Animals like fish or reptiles usually combine the fact that their eyes are placed on the sides of the head (and thus each of them has limited access to hemifields of space accessible to the contralateral eye) together with very reduced ipsilateral projections. They have no structure homologous to the corpus callosum, instead possessing only a small anterior commissure and a small interhemispheric hippocampal commissure that interconnects regions of the dorsal part of the telencephalon. Neuroanatomically, they can be considered quite close to "split brain" preparations (see Deckel, 1995, 1997).

Deckel's work is on *Anolis* lizards, and he frequently compares them to the split-brain condition: "Unlike mammals, the visual system of *Anolis* can, in some ways, be considered as a 'split brain' preparation, i.e., a condition where the left hemisphere of the brain is relatively 'unaware' of information perceived and processed by the right hemisphere of the brain." A. Wallace Deckel, "Laterality of Aggressive Responses in *Anolis*," *Journal of Experimental Zoology* 272, no. 3 (1995): 194–200.

240 *The blind cave fish have the same preference*: See Theresa Burt de Perera and Victoria A. Braithwaite, "Laterality in a Non-Visual Sensory Modality—The Lateral Line of Fish," *Current Biology* 15, no. 7 (2005): R241–42. Victoria Braithwaite died

during the later stages of the writing of this book. She did terrific work on fish sentience, and though I only knew her slightly, she seemed like a wonderful person.

240 *sometimes just half the animal goes pale, and the other half does not*: The fleeing motion itself, though, seems to be immediately coordinated across the whole body; it's not just the signaling side whose arms prepare for escape. In some cases the escape is by a simple jet with the arms relaxed and trailing—easy to do. But in other cases the escape is an eight-armed clamber. (The book *Other Minds* has a drawing of this split color, in the last chapter.)

240 *Experiments have shown in some cases that if a bird learns a task*: See Laura Jiménez Ortega et al., "Limits of Intraocular and Interocular Transfer in Pigeons," *Behavioural Brain Research* 193, no. 1 (2008): 69–78.

240 *I was startled to learn that no corpus callosum is present in marsupials*: See Rodrigo Suárez et al., "A Pan-Mammalian Map of Interhemispheric Brain Connections Predates the Evolution of the Corpus Callosum," *Proceedings of the National Academy of Sciences USA* 115, no. 38 (2018): 9622–27. They say that "humans with congenital absence of the corpus callosum, but with preserved interhemispheric integrative functions, often show compensatory wiring through the anterior commissure that resembles the noneutherian connectome."

242 *Giorgio Vallortigara, whose work with chicks I mentioned earlier*: This is in his "Comparative Neuropsychology of the Dual Brain."

242 *The extent of the differences will depend on questions*: Earlier I also compared the split-brain situation to the Wada procedure, in which each half of the upper brain is put to sleep in a patient in turn. The Wada procedure was used in that earlier discussion to support the possibility of fast switching in split-brain patients. In a Wada procedure, because the corpus callosum is intact, both of the halves within the procedure, and also the whole brain, outside the procedure, have the kinds of internal connectedness that enable large-scale dynamic properties to be unified. In the fast switching that would occur in split-brain cases, the "one mind" condition would be one in which a lot of the connections that integrate large-scale dynamic patterns would be absent.

244 *Dolphins have extremely large brains*: See Kieran C. R. Fox, Michael Muthukrishna, and Susanne Shultz, "The Social and Cultural Roots of Whale and Dolphin Brains," *Nature Ecology & Evolution* 1 (2017): 1699–705; Lori Marino, Daniel W. McShea, and Mark D. Uhen, "Origin and Evolution of Large Brains in Toothed Whales," *The Anatomical Record Part A, Discoveries in Molecular Cellular and Evolutionary Biology* 281, no. 2 (2004): 1247–55; and Richard C. Connor, "Dolphin Social Intelligence: Complex Alliance Relationships in Bottlenose Dolphins and a Consideration of Selective Environments for Extreme Brain Size Evolution in Mammals," *Philosophical Transactions of the Royal Society of London B, Biological Sciences* 362 (2007): 587–602.

244 *As eutherian mammals, dolphins have a corpus callosum*: See Raymond J. Tarpley and Sam H. Ridgway, "Corpus Callosum Size in Delphinid Cetaceans," *Brain, Behavior and Evolution* 44, no. 3 (1994): 156–65.

245 *I don't know why she singled him out*: Was it his red hair? Dolphins are supposed to be color-blind, unlike their terrestrial relatives. "Supposed to be"? Their eye physiology suggests they should be, but there may be hidden means for color discrimination in dolphins, as in octopuses. The behavioral evidence seems ambiguous; see Ulrike Griebel and Axel Schmid, "Spectral Sensitivity and Color Vision

in the Bottlenose Dolphin (*Tursiops truncatus*)," *Marine and Freshwater Behaviour and Physiology* 35, no. 3 (2002): 129–37. They found some color sensitivity, in one animal. Some interesting recent ideas about color sensitivity in octopuses may also apply to dolphins, as the authors note; see Alexander L. Stubbs and Christopher W. Stubbs, "Spectral Discrimination in Color Blind Animals via Chromatic Aberration and Pupil Shape," *Proceedings of the National Academy of Sciences USA* 113, no. 29 (2016): 8206–11.

246 *The land has a third of Earth's area but about 85 percent of its species*: See Geerat J. Vermeij and Richard K. Grosberg, "The Great Divergence: When Did Diversity on Land Exceed That in the Sea?," *Integrative and Comparative Biology* 50, no. 4 (2010): 675–82; and Grosberg, Vermeij, and Peter C. Wainwright, "Biodiversity in Water and on Land," *Current Biology* 22, no. 21 (2012): R900–903.

247 *In a 2017 paper, Vermeij argues for this by going through a list of evolutionary innovations*: This is his "How the Land Became the Locus of Major Evolutionary Innovations," *Current Biology* 27, no. 20 (2017): 3178–82.

10. PUT TOGETHER BY DEGREES

249 *When I awoke at midnight, not knowing where I was*: From Proust's *Swann's Way*, trans. C. K. Scott Moncrieff (New York: Henry Holt, 1922).

251 *The Canadian psychologist Endel Tulving, who named episodic memory*: The original paper is his "Episodic and semantic memory," in Endel Tulving and Wayne Donaldson, *Organization of Memory* (New York: Academic Press, 1972). The patient Kent Cochrane was originally known just as "KC."

251 *Clive Wearing, an English expert on early music*: His wife, Deborah Wearing, wrote a book about his case and their life together, called *Forever Today: A Memoir of Love and Amnesia* (London: Transworld, 2004). Wearing is also described in Oliver Sacks's "The Abyss," *The New Yorker*, September 24, 2007.

251 *Around 2007 a series of papers appeared*: These include Donna Rose Addis, Alana T. Wong, and Daniel L. Schacter, "Remembering the Past and Imagining the Future: Common and Distinct Neural Substrates During Event Construction and Elaboration," *Neuropsychologia* 45, no. 7 (2007): 1363–77, and Demis Hassabis, Dharshan Kumaran, and Eleanor A. Maguire, "Using Imagination to Understand the Neural Basis of Episodic Memory," *Journal of Neuroscience* 27, no. 52 (2007): 14365–74. I rely here especially on Thomas Suddendorf, Donna Rose Addis, and Michael C. Corballis, "Mental Time Travel and the Shaping of the Human Mind," *Philosophical Transactions of the Royal Society B* 364 (2009): 1317–24; Daniel L. Schacter et al., "The Future of Memory: Remembering, Imagining, and the Brain," *Neuron* 76, no. 4 (2012): 644–94; also Donna Rose Addis, "Are Episodic Memories Special? On the Sameness of Remembered and Imagined Event Simulation," *Journal of the Royal Society of New Zealand* 48, no. 2–3 (2018): 64–88. The point about remembering from unexperienced visual perspectives is made in Schacter and Addis, "Memory and Imagination: Perspectives on Constructive Episodic Simulation," forthcoming in *The Cambridge Handbook of the Imagination*, ed. Anna Abraham (Cambridge, UK: Cambridge University Press, 2020).

253 *Donna Rose Addis, one of the leading psychologists in this area*: This is seen in "Are Episodic Memories Special?"

254 *the first detailed theory of dreams with a good basis in neurobiology (I do not include Freud) was developed by the Harvard psychiatrists Allan Hobson and Robert McCarley*:

NOTES

"The Brain as a Dream State Generator: An Activation-Synthesis Hypothesis of the Dream Process," *American Journal of Psychiatry* 134, no. 12 (1977): 1335–48. My discussion of dreams in this section draws extensively on Erin J. Wamsley and Robert Stickgold, "Dreaming and Offline Memory Processing," *Current Biology* 20, no. 23 (2010): R1010–13.

254 *Francis Crick and Graeme Mitchison later hypothesized that dreams are junk*: "The Function of Dream Sleep," *Nature* 304 (1983): 111–14.

254 *Against this background, some recent theories of dreams are the most sensible-seeming*: Here I am drawing on Wamsley and Stickgold, "Dreaming and Offline Memory Processing."

255 *In an evocative phrase lurking around philosophy since the time of Martin Heidegger*: The phrase is one translation of the German *Dasein*, which is central to Martin Heidegger's *Being and Time* (1927), though apparently Heidegger did not think that "being there" was a good translation of his *Dasein* (see Hubert L. Dreyfus, *Being-in-the-World: A Commentary on Heidegger's "Being and Time"* [Cambridge, MA: MIT Press, 1990]). Andy Clark's book is *Being There: Putting Brain, Body, and World Together Again* (MIT Press, 1997). Hegel had used the term *Dasein* also, in his *Science of Logic* (1812–16), but Heidegger made it famous.

255 *Sleep itself is extremely widespread in animals*: For a review, see Alex C. Keene and Erik R. Duboue, "The Origins and Evolution of Sleep," *Journal of Experimental Biology* 221, no. 11 (2018): jeb159533. Even some jellyfish (Cubozoa, box jellyfish) have something like sleep.

255 *Cuttlefish, which are octopus relatives but often even more colorful, have been the subjects of two remarkable studies of sleep*: Marcos G. Frank et al., "A Preliminary Analysis of Sleep-Like States in the Cuttlefish *Sepia officinalis*," *PLOS ONE* 7, no. 6 (2012): e38125, and Teresa L. Iglesias et al., "Cyclic Nature of the REM Sleep–Like State in the Cuttlefish *Sepia officinalis*," *Journal of Experimental Biology* 222 (2019): jeb174862.

256 *"Place cells" are part of a well-studied system*: This is a huge literature. A classic is John O'Keefe and Lynn Nadel, *The Hippocampus as a Cognitive Map* (Oxford, UK: Clarendon / Oxford University Press, 1978), and for some recent work, H. Freyja Ólafsdóttir et al., "Hippocampal Place Cells Construct Reward Related Sequences Through Unexplored Space," *eLife* 4 (2015): e06063, and H. Freyja Ólafsdóttir, Daniel Bush, and Caswell Barry, "The Role of Hippocampal Replay in Memory and Planning," *Current Biology* 28, no. 1 (2018): R37–50. Recent work has been able to show an ongoing interaction between replay and preplay in awake rats, as they learn and plan: see Justin D. Shin, Wenbo Tang, and Shantanu P. Jadhav, "Dynamics of Awake Hippocampal-Prefrontal Replay for Spatial Learning and Memory-Guided Decision Making," *Neuron* 104, no. 6 (2019): 1110–25.e7.

257 *The replay of paths can be compared within REM and slow-wave sleep*: For slow-wave, see Thomas J. Davidson, Fabian Kloosterman, and Matthew A. Wilson, "Hippocampal Replay of Extended Experience," *Neuron* 63, no. 4 (2009): 497–507; for replay at a natural pace in REM sleep, Kenway Louie and Matthew A. Wilson, "Temporally Structured Replay of Awake Hippocampal Ensemble Activity during Rapid Eye Movement Sleep," *Neuron* 29, no. 1 (2001): 145–56. The comparison is discussed in Ólafsdóttir, Bush, and Barry, "The Role of Hippocampal Replay in Memory and Planning."

261 *If you are a living thing of the right kind (and awake, and so on), you have an*

317

experiential profile: Although this is intended as a weaker claim (for now), I have some sympathy with Michael Tye's "one experience" view: see his *Consciousness and Persons* (Cambridge, MA: MIT Press, 2003). Something that concerns me is squaring this view with ideas about disunity and partial unity defended in Chapters 6 and 8. This notion of an experiential profile is also reminiscent of Searle's "unified field" view of consciousness, but I think that view might be one of those misuses of the "field" idea that is discussed in this chapter and Chapter 7. See his "Consciousness," 2000.

262 *The lights in an animal (ant, jellyfish, squid) are either on or not*: This was discussed at an NYU Animal Consciousness conference in 2006. Michael Tye rejected gradualism in his talk and discussion, as did some others. See also Jonathan A. Simon, "Vagueness and Zombies: Why 'Phenomenally Conscious' Has No Borderline Cases," *Philosophical Studies* 174 (2017): 2105–23. Tim Bayne, Jakob Hohwy, and Adrian M. Owen say that "the notion of degrees of consciousness is of dubious coherence": "Are There Levels of Consciousness?," *Trends in Cognitive Sciences* 20, no. 6 (2016): 405–13.

264 *I suppose they would say the same thing about a human first acquiring consciousness*: See Alison Gopnik, *The Philosophical Baby: What Children's Minds Tell Us About Truth, Love, and the Meaning of Life* (New York: Farrar, Straus and Giroux, 2009).

265 *In much current work, people try to explain conscious experience with a detailed information-processing model*: Stanislas Dehaene's version of the "workspace" view is my focus here; see his *Consciousness and the Brain: Deciphering How the Brain Codes Our Thoughts* (New York: Viking, 2014). Other views in the same family include Prinz's AIR view, spelled out in *The Conscious Brain* (Oxford, UK: Oxford University Press, 2012), and Michael Tye's PANIC view, in *Ten Problems of Consciousness: A Representational Theory of the Phenomenal Mind* (Cambridge, MA: MIT Press, 1995).

266 *I respect those experimental findings, but I disagree with how they are often interpreted*: Morten Overgaard's work is a helpful stimulus to alternative interpretations; see, for example, Morten Overgaard et al., "Is Conscious Perception Gradual or Dichotomous? A Comparison of Report Methodologies During a Visual Task," *Consciousness and Cognition* 15 (2006): 700–708.

266 *The French neuroscientist Stanislas Dehaene*: See *Consciousness and the Brain*.

272 *In the area of AI, then, the biggest disagreement I have is with "upload" scenarios*: For a collection of papers about uploading, see Russell Blackford and Damien Broderick, eds., *Intelligence Unbound: The Future of Uploaded and Machine Minds* (Malden, MA: John Wiley and Sons, 2014), including the pair of papers by David Chalmers and Massimo Pigliucci.

273 *Some recent proposals have said that once there is a reasonable chance that an animal is sentient*: See Jonathan Birch, "Animal Sentience and the Precautionary Principle," *Animal Sentience* 2017.017.

274 *This book has not included a lot of material about mammals*: Hobson's work on dreams, mentioned earlier in this chapter, included very disturbing experiments on cats.

274 *Questions about how the brain works are being transformed by that research*: See the work on "place cells" cited above. Nobel Prizes were awarded in 2014 to John O'Keefe (half) and May-Britt Moser and Edvard Moser (half, jointly). For a discussion of animal experimentation with some specific attention to the case of rats,

which are used now in massive numbers, see Phillip Kitcher, "Experimental Animals," *Philosophy and Public Affairs* 43, no. 4 (2015): 287–311.

275 *Even when reasoning of this kind dominates a situation, much can be done*: For all these issues, see Lori Gruen, *Ethics and Animals: An Introduction* (Cambridge, UK: Cambridge University Press, 2011).

278 *The term was invented by Haeckel*: It appears in his "Our Monism" (1892); I discuss the idea in "Mind, Matter, and Metabolism," *Journal of Philosophy* 113, no. 10 (2016): 481–506.

280 *The philosopher John Dewey, back in the 1920s*: See his *Experience and Nature* (Chicago: Open Court, 1925), 227–28.

280 *Thirty years ago, as a student, I went to a conference*: This was the 1989 Eastern Division Meeting of the American Philosophical Association, which had a symposium called "Thought of Wittgenstein," with Crispin Wright, Warren Goldfarb, and John McDowell.

280 *Wittgenstein, working in the early and mid-twentieth century, is the person who argued*: See his *Philosophical Investigations*, trans. G.E.M. Anscombe (London: Basil Blackwell, 1953), and Gilbert Ryle's *The Concept of Mind* (Chicago: University of Chicago Press, 1949). Here is Ryle: "The statement 'the mind is its own place,' as theorists might construe it, is not true, for the mind is not even a metaphorical 'place.' On the contrary, the chessboard, the platform, the scholar's desk, the judge's bench, the lorry-driver's seat, the studio and the football field are among its places. These are where people work and play stupidly or intelligently."

281 *Following this theme, one speaker at the meeting (Crispin Wright)*: I am pretty sure it was Wright who introduced the idea, which was also discussed by others. See Wright's "Wittgenstein's Later Philosophy of Mind: Sensation, Privacy, and Intention," in *Meaning Scepticism*, ed. Klaus Puhl (Berlin: De Gruyter, 1991), 126–47 (not Wright's paper with this title in the *Journal of Philosophy*) and McDowell's "Intentionality and Interiority in Wittgenstein," in the same collection.

ACKNOWLEDGMENTS

As this book grew, it took on a tentacular form that wandered into more and more animal groups, an ever-increasing list of research literatures, and countless other scientific and philosophical paths and alleys. I would have been entirely lost at many stages without the generous help of discussants and correspondents with expertise in these areas. Listing people in rough order of chapters, I want to thank Chris Shields, Alison Simmons, Gary Hatfield, Maureen O'Malley, Gáspár Jékely, Tom Davis, Dave Harasti, Meryl Larkin, Jim Gehling, John Allen, Steve Whalen, Gary Cobb, Andrew Barron, Pam Lyon, Nick Lane, Derek Denton, Björn Merker, Björn Brembs, Madeleine Beekman, Kim Sterelny, Andrew Knoll, Nick Strausfeld, Jonathan Birch, Evan Thompson, Michael Kuba, Elizabeth Schechter, Tim Bayne, Bruno van Swinderen, Lars Chittka, Cwyn Solvi, Claire O'Callaghan, Christof Koch, Terry Walters, Katherine Preston, Monica Gagliano, and Lesley Rogers. Particular thanks to Rosa Cao, who, together with Fred Keijzer, nudged me toward becoming a less orthodox philosopher of mind, and to Lori Gruen, who long ago led me to see animals differently.

I am delighted to include the work of two illustrators, Alberto Rava and Rebecca Gelernter. Alberto is a diver and an artist with a flair for capturing how animals move and inhabit their bodies; his work appears on pages 53, 63, 89, 142, 177, 193, 210, and 234. Rebecca

ACKNOWLEDGMENTS

Gelernter's characteristic combination of elegance and scientific accuracy can be seen on pages 45, 61, 69, 82, 98, 126, and 237.

The copy editor's role often remains silent in acknowledgments, but for the second time Annie Gottlieb has done an excellent job, adding subtle but substantial improvements. On the marine side, many thanks to Matt Lawrence, David Scheel, Marty Hing, and Kylie Brown. Thanks to Jim Ris Mamuko, my dive guide at Kungkungan Bay Resort in Lembeh, and Chris Jansen and Katie Anderson at Live Ningaloo for much help with whale sharks. Thanks also to Mick Todd at Let's Go Adventures for his magic pudding air-fills and more, to Trudie Campey at Feet First Dive, and to Richard Nicholls at Dive Center Manly. I again thank the custodians of Cabbage Tree Bay Aquatic Reserve, Booderee National Park, Jervis Bay Marine Park, and Port Stephens–Great Lakes Marine Park for their work in caring for these extraordinary places.

Along with the people who helped me follow tentacular scientific paths, and those in and around the water, are a few people who were involved in every part and every stage. They are the brilliant Jane Sheldon, whom I thank for so many deft formulations and the sharpest critical eye; Alex Star, my editor, who helped shape this material, at all scales, with great care and insight; and my agent, Sarah Chalfant, whose role has been transformative of the entire project.

INDEX

Page numbers in *italics* refer to illustrations.

A NOTE ABOUT THE AUTHOR

Peter Godfrey-Smith is a professor in the School of History and Philosophy of Science at the University of Sydney. He is the author of *Other Minds: The Octopus, the Sea, and the Deep Origins of Consciousness* as well as *Theory and Reality: An Introduction to the Philosophy of Science* and *Darwinian Populations and Natural Selection*, which won the 2010 Lakatos Award.